THE ORIGIN OF THE SOLAR SYSTEM
The Capture Theory

**THE ELLIS HORWOOD LIBRARY OF SPACE
SCIENCE AND SPACE TECHNOLOGY**
SERIES IN ASTRONOMY

Series Editor: JOHN MASON
Consultant Editor: PATRICK MOORE

This series aims to coordinate a team of international authors of the highest reputation, integrity and expertise in all aspects of astronomy. It will make a valuable contribution to the existing literature encompassing all areas of astronomical research. The titles will be illustrated with both black and white and colour photographs, and include many line drawings and diagrams, with tabular data and extensive bibliographies. Aimed at a wide readership, the books will appeal to the professional astronomer, undergraduate students, the high-flying 'A' level student, and the non-scientist with a keen interest in astronomy.

PLANETARY VOLCANISM*
PETER CATTERMOLE, Department of Geology, Sheffield University, UK
SATELLITE ASTRONOMY: The Principles and Practice of Astronomy from Space
JOHN K. DAVIES, Royal Observatory, Edinburgh, UK
THE DUSTY UNIVERSE*
ANEURIN EVANS, Department of Physics, University of Keele, UK
SPACE-TIME AND THEORETICAL COSMOLOGY*
MICHEL HELLER, Department of Philosophy, University of Cracow, Poland
ASTEROIDS: Their Nature and Utilization
CHARLES T. KOWAL, Space Telescope Institute, Baltimore, Maryland, USA
ELECTRONIC AND COMPUTER-AIDED ASTRONOMY*
IAN S. McLEAN, Joint Astronomy Centre, Hilo, Hawaii, USA
URANUS: The Planet, Rings and Satellites*
ELLIS D. MINER, Jet Propulsion Laboratory, Pasadena, California, USA
THE PLANET NEPTUNE
PATRICK MOORE
ACTIVE GALACTIC NUCLEI*
IAN ROBSON, Director of Observatories, Lancashire Polytechnic, Preston, UK
ASTRONOMICAL OBSERVATIONS FROM THE ANCIENT ORIENT*
RICHARD F. STEPHENSON, Department of Physics, Durham University, Durham, UK
**SURFACE GEOLOGY OF TERRESTRIAL PLANETS AND SATELLITES: Instrumentation,
Investigation, Interpretation**
YURI A. SURKOV, Chief of the Laboratory of Geochemistry of the Planets, USSR Academy of Sciences, Moscow, USSR
THE HIDDEN UNIVERSE*
ROGER J. TAYLER, Astronomy Centre, University of Sussex, UK
THE ORIGIN OF THE SOLAR SYSTEM: The Capture Theory
JOHN R. DORMAND, Department of Mathematics and Statistics, Teesside Polytechnic, Middlesborough, UK, and
MICHAEL M. WOOLFSON, Department of Physics, University of York, UK
AT THE EDGE OF THE UNIVERSE*
ALAN WRIGHT, Australian National Radio Astronomy Observatory, Parkes, New South Wales, and HILARY WRIGHT

* *In preparation*

THE ORIGIN OF THE SOLAR SYSTEM

The Capture Theory

JOHN R. DORMAND, B.Sc., D.Phil.
Department of Mathematics & Statistics
Teesside Polytechnic

MICHAEL M. WOOLFSON, FRS, F.Inst.P., Ph.D.
Professor of Theoretical Physics
University of York

ELLIS HORWOOD LIMITED
Publishers · Chichester

Halsted Press: a division of
JOHN WILEY & SONS
New York · Chichester · Brisbane · Toronto

First published in 1989 by
ELLIS HORWOOD LIMITED
Market Cross House, Cooper Street,
Chichester, West Sussex, PO19 1EB, England
The publisher's colophon is reproduced from James Gillison's drawing of the ancient Market Cross, Chichester.

Distributors:

Australia and New Zealand:
JACARANDA WILEY LIMITED
GPO Box 859, Brisbane, Queensland 4001, Australia

Canada:
JOHN WILEY & SONS CANADA LIMITED
22 Worcester Road, Rexdale, Ontario, Canada

Europe and Africa:
JOHN WILEY & SONS LIMITED
Baffins Lane, Chichester, West Sussex, England

North and South America and the rest of the world:
Halsted Press: a division of
JOHN WILEY & SONS
605 Third Avenue, New York, NY 10158, USA

South-East Asia
JOHN WILEY & SONS (SEA) PTE LIMITED
37 Jalan Pemimpin # 05–04
Block B, Union Industrial Building, Singapore 2057

Indian Subcontinent
WILEY EASTERN LIMITED
4835/24 Ansari Road
Daryaganj, New Delhi 110002, India

© **1989 J.R. Dormand and M.M. Woolfson/Ellis Horwood Limited**

British Library Cataloguing in Publication Data
Dormand, John R.
The origin of the solar system: the Capture theory.
1. Solar system. Origins
I. Title   II. Woolfson, Michael M.
521′.54

Library of Congress data available

ISBN 0–7458–0601–5 (Ellis Horwood Limited)
ISBN 0–470–21466–X (Halsted Press)

Typeset in Times by Ellis Horwood Limited

Printed and bound in Great Britain at
The Camelot Press Ltd, Southampton
COPYRIGHT NOTICE

Table of contents

Preface

It is a quarter of a century since the Capture Theory of the origin of the solar system was first published. During this period our knowledge of the physical and dynamical properties of the solar system has increased dramatically mainly due to the applications of space technology. Consequently it would not cause much surprise if existing theories of cosmogony needed much revision in the light of new data. Yet this has not been the case with the Capture Theory. Much of the new knowledge, highlighting the irregular and chaotic aspects of our system, lends support to the capture hypothesis, and there is a new realization that catastrophic events, at one time unfashionable and often dismissed as too 'unlikely' to contemplate, must have played an important role in shaping the planets and satellites. Some other cosmogonic ideas, relying totally on what might be described as 'regular' processes, must now be relegated to the history of astronomy. In case of misunderstanding it must be emphasized here that the terms 'regular' and 'irregular' do not apply to the application of physical laws. rather they are applied to the types of processes which may occur in celestial systems. For example, the earth's orbital motion may be classified as regular while the ancient bombardment of the moon is in the other category. The former can be predicted accurately in a completely deterministic way but the latter is a stochastic process. Nevertheless an actual impact on the moon of a massive projectile may be described in the terms of conventional physics.

For some time now we have felt that it would be useful to gather together in a single volume many of the ideas which comprise the Capture Theory of cosmogony. These ideas have long been available in the scientific journals but these are really only accessible to specialists, and so, in addition to providing an overall picture of the theory, we have aimed this book primarily at the reader who has a basic knowledge of astronomy and physics, although we hope that it may be of interest and use to a wider audience. We have tried to make the subject of cosmogony as accessible as possible while preserving a reasonable degree of detail. The inclusion of many figures relating to aspects of dynamical astronomy will help the inexperienced reader, and perhaps serve to consolidate an understanding of the subject for those who have been introduced to it previously.

It is impossible to ignore the fact that astronomy is a quantitative science. The application of mathematical techniques is essential to all astronomical endeavours, particularly the study of cosmogony, and the Capture Theory has been subjected to a great deal of numerical modelling for the purpose of computer simulation, without which no modern theory would be complete. Many of the results and figures to be found in this book derive from such computer analysis but a full description of the methods employed is not presented for the reasons given earlier. Nevertheless some of the more basic mathematical formulae and equations do appear in the relevant chapters.

Most of this volume is concerned with the Capture Theory but no book on cosmogony should ignore the many differing opinions on this subject. Thus a review of modern and also older discredited theories is presented. It will become clear that the authors feel that the Capture Theory is much more successful than any other in explaining the known features of the solar system. Other astronomers may not agree that this is the case, but we must accept that planetary cosmogony will always be the subject of proper scientific debate; even eminent scientists will often differ in their judgements of the correctness of theories.

In producing this book we have had a great deal of help, particularly with the figures and plates. We especially thank Mr Alan Gebbie of the Physics Department at York for his clear artwork which assists so much in the explanation of the science. We are also grateful to the Jet Propulsion Laboratory and to Dr R. Hutchison of the British Museum for providing material for the plates.

1

First ideas

1.1 THE EMERGENCE OF THE PROBLEM

A trait which man shares with many of his fellow creatures is an intense curiosity, but only in man is that curiosity coupled with high intelligence and boundless imagination. It is not surprising therefore to find that even the most primitive societies have speculated on the origin of the earth and its relationship to the sun, the moon and the stars. The first verse of the first chapter of Genesis is concerned with these very matters and many ancient cultures directly associated the earth and various heavenly bodies with deities.

The science of astonomy exercised many of the greatest minds of pre-Christian Greece. From this period there arises the concept of a spherical earth (Pythagoras, 572–492 B.C.) and the explanation of eclipses (Empedocles, 484–424 B.C.). Measurements were made on the relative sizes of the earth, moon and sun and on the distances of the moon and the sun from the earth (Aristarchus, 310–230 B.C.) and an absolute size for the earth was found by Eratosthenes (276–195 B.C.).

The problem of the relative motion of the sun, the earth and the planets also stimulated much speculation. The pressure for believing in an earth-centred universe was very great; firstly there is the collective egocentricity of man — everything exists for his benefit — and secondly the complete lack of sensation of motion for the earth-dweller. It should be recalled that personal movement for ancient man involved walking, riding a horse, being pulled in a carriage or travelling in a ship, for all of which the sensation of motion is evident, sometimes uncomfortably so. There were those who hinted at a heliocentric system and, indeed, Aristarchus was quite explicit in putting forward this model. However, such a view did not become established and Ptolemy's earth-centred model, formulated in the second century A.D., with its complicated combination of conventional circular orbits and epicycles, became the established truth for the next 1300 years or so.

Nicholas Copernicus was born in Torun, Poland, in 1473. He was destined to become a church administrator and his education spanned the complete range of knowledge of his time. He became interested in astronomy, made measurements of planetary motions, and through these became certain that the motion of the planets could be reasonably explained only by a heliocentric theory. He set down his ideas in a book *De revolutionibus orbium coelestium*, dedicated to Pope Paul III, which was published shortly before his death in 1543. For a few years the Church seemed to support, or at least to tolerate, this work but, in time, its attitude changed. The realization gradually dawned that, by demoting the earth, man himself was being demoted and that this new theory might be the thin edge of a wedge in an attack on the Church and even on God's role in the universe.

The century following the death of Copernicus was dominated by the works of two great astronomical observers. The first of these was Tycho Brahe (1546–1601), a Danish nobleman, who built two magnificent observatories equipped with accurate line-of-sight instruments. His precise and numerous observations of planetary motion were compiled into tables which were of inestimable value to later astronomers. Nevertheless, for all the precision of his measurements, Tycho Brahe remained convinced that the earth was at the centre of the system consisting of the planets and the sun.

The contribution of the second great observer, Galileo (1564–1642), was of a different kind. His improved version of the recently invented telescope allowed him to observe features of the solar system that were previously invisible, e.g. the four largest satellites of Jupiter and the phases of Venus. Galileo was led by his observations and by his intellect to espousing the Copernican theory and this brought him, most reluctantly, into a head-on collision with the Church. He was forced by physical threats to recant publicly his belief in the heliocentric theory, but there can be little doubt that he believed it even so.

During the last few years of his life Tycho Brahe worked in Prague under the patronage of Rudolph II of Bohemia. In 1599 he was assisted in preparing his tables by Johannes Kepler (1571–1630) who supported the Copernican view of the solar system. From an analysis of Tycho Brahe's accurate observations Kepler was led to his three famous laws of planetary motion. These were:

(1) Planets move in elliptic orbits with the sun at one focus.
(2) The line joining a planet to the sun sweeps out equal areas in equal times.
(3) The square of the orbital period is proportional to the cube of the average distance from the sun.

These laws were purely empirical and Kepler had no idea of the underlying principles which governed them. For these principles the world had to wait but a few years for the birth of the greatest scientific genius of all time.

Isaac Newton (1642–1727) was born in the year of Galileo's death. The extent of his contributions to mathematics and science is unparalleled but of all these we note only his explanation of Kepler's three laws in terms of an inverse-square law of gravitational attraction between bodies. This was the final and incontrovertible step in establishing the validity of the Copernican system. Albeit that *action at a distance* was, and still is, a difficult concept, Newton's work established a proper understanding of the dynamics of the solar system and provided a sound base from which to consider theories of its origin.

The planets known at the beginning of the eighteenth century were those out to Saturn, which could be seen easily with the naked eye and were also known to the ancients. In 1772 Johann Elert Bode (1747–1826), probably influenced by an earlier work of Johann Titius von Wittenberg, noted that the distances of the planets from the sun seemed to follow a simple mathematical progression, as is shown in Table 1.1. True there was a gap between Mars and Jupiter but otherwise the law held

Table 1.1 — Bode's law gives the orbital radius of Mercury as 0.4 astronomical units (AU). For the other planets the orbital radii are $0.4+0.3\times2n$ AU with $n = 0$ for Venus, $n = 1$ for the earth, etc

Planet	Mean orbital radius (AU)	Bode's law value
Mercury	0.39	0.4
Venus	0.72	0.7
Earth	1.00	1.0
Mars	1.52	1.6
Ceres	2.8	2.8
Jupiter	5.20	5.2
Saturn	9.54	10.0
Uranus	19.19	19.6

remarkably well. When in 1781 Herchel discovered the planet Uranus its distance was found to fit Bode's law and the law was given further reinforcement in 1801 with the discovery by Piazzi of Ceres, the largest asteroid, which occupies the gap between Mars and Jupiter.

The pattern which emerges at the end of the eighteenth century is of a regular, almost planar, solar system with planets in well-ordered orbits all rotating in the same sense. The largest planets possessed substantial satellite systems which resembled miniature versions of the planetary system. That which is ordered and regular should be explicable, and in the eighteenth century there began an onslaught of scientific study and analysis of the problem of the origin of the solar system which has not ceased to the present day.

1.2   THE NEBULA THEORIES OF LAPLACE AND ROCHE

During the eighteenth century observational astronomers were discovering nebulae and this led the philosopher Kant to propose that stars originated as condensations of such clouds. He also proposed that the residue of the cloud, after the central star had formed, could give the material for the formation of planets. The same idea was given, probably independently and in a much improved form, by Pierre Laplace in his *Exposition du système du monde* first published in 1796.

Laplace argued that a nebula, by virtue of radiating energy and so becoming cooler, would gradually collapse inwards under self-gravitational forces. In doing so, in order to conserve angular momentum, it would rotate more rapidly and so flatten along its rotational axis, eventually taking on a lenticular shape. At this stage the equatorial speed of the nebula would correspond to orbital motion around the central mass and, thereafter, orbiting matter would be left behind in the equatorial plane. Laplace postulated that the orbiting material would separate from the central mass in a spasmodic way giving rise to a series of orbiting rings. He then assumed that material in each ring would collect together to form agglomerations and that one of these in each ring would be massive enough to attract and absorb all the others to form a single planet. The various stages in this process are shown in Fig. 1.1. Finally

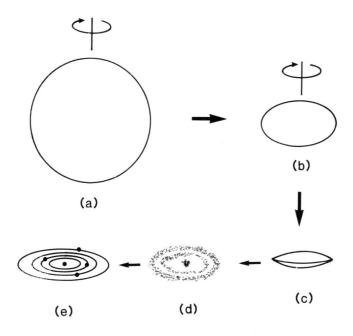

Fig. 1.1 — (a) A rotating nebula. (b) The collapsing nebula flattened along its rotation axis. (c) Formation of a lenticular shape. (d) A series of rings left behind by the contracting core. (e) One residual condensation in each ring forms a planet.

Laplace proposed that satellites arose in much the same way from condensing rotating protoplanets and he considered the rings of Saturn to be those which had not yet gone through the process of forming satellites.

There is a fundamental and insuperable difficulty with the model as described. A striking characteristic of the solar system is that the planets, with about 1/700 of the mass of the system, in their orbital motion account for over 99% of its angular momentum. There seems to be no way in which an initially diffuse nebula could evolve so as to partition mass and angular momentum in that way. It turns out, as we shall see, that the angular momentum problem is one of the most important hurdles to be negotiated by any plausible theory for the origin of the solar system.

An attempt to overcome this problem was made by Roche in 1854. A rigidly rotating nebula was imagined to be enveloping the sun which already existed as a highly condensed object. The nebula, which developed in the way proposed by Laplace, had only to provide the planets, the formation of which was decoupled from that of the sun.

Theoretical astronomers frequently work with idealized models and a gaseous cloud or star with all the mass concentrated at the centre and with an envelope of finite extent, but negligible mass, is known as a Roche model. In 1919 Jeans studied the behaviour of such a model when it was rotating and collapsing. He found that for the outer material to be able to condense, as it must to form a planet, its density must be greater than 0.361 times the mean density of the whole system, including the contribution of the central mass. This would imply that the envelope had a mass of similar magnitude to that of the central body — which is inconsistent with the Roche model — and all the difficulties of the Laplace hypothesis remain. However, as we shall see later, ideas for reviving the nebula theory have persisted even to the recent past.

1.3   TWO-BODY INTERACTION THEORIES

As early as 1745 it was suggested by Buffon that material for the planets was knocked out of the sun by a grazing interaction of a comet with the sun's surface. This material was then supposed to have formed itself into planets at various distances from the sun. In his *Exposition du système du monde* Laplace criticized Buffon's hypothesis on the ground that the orbits of portions of matter ejected from the sun would return to the sun's surface and so be reabsorbed. He then conceded that the mutual interactions of different parts of the ejected material might modify the orbits and negate his argument, but he concluded that, in any case, the very eccentric orbits of the resultant planets would be at variance with observation. Laplace was convinced that initially near-circular orbits were an essential outcome for a correct theory and this was so for his own nebula hypothesis.

Chamberlin, in 1901, and Moulton, in 1905, proposed another kind of two-body theory which involved the interaction of a massive star with the sun. These workers assumed that during a period when there were large and periodic eruptions from the sun (prominences) a massive star passed so close to the sun that the ejected material was drawn out further, then attracted by the retreating star and left in orbit around the sun. Matter would have been drawn out of the sun from those regions where there were the greatest tidal effects; that is, the part of the sun's surface closest to the massive star and also the part which was most remote; and the ejected matter would have resembled the arms of a spiral nebula (Fig. 1.2). It was suggested that the matter in these arms would have condensed to form small bodies, called planetesimals, and these would have been bunched together, each bunch corresponding to a different prominence. The aggregation of a group of planetesimals would form a planet; it was also assumed that a large prominence would have associated with it smaller subsidiary eruptions of matter and that these would have given rise to satellite families. The theory, as given, was descriptive and devoid of mathematical treatment. It was an analysis of this theory by Jeans which led him, in 1917, and later Jeffreys, in 1918, to the development of the important tidal theory which, for some time, was accepted as having solved the problems of the origin of the solar system.

The general form of Jeans tidal theory is illustrated in Fig. 1.3. Jeans found that

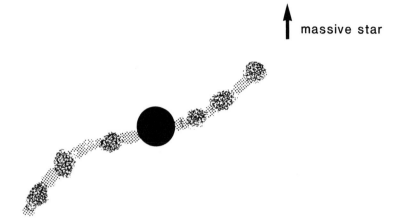

Fig. 1.2 — The Chamberlin and Moulton theory. Each bunched region results from a different solar prominence.

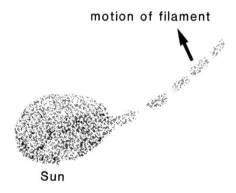

Fig. 1.3 — A general view of the Jeans tidal theory.

tidal action alone, without the need of prominences, could cause material to be drawn out of the sun by a passing massive star. Due to the nearness of the star there would be a dominant tidal bulge on the side of the sun facing the star and, when the star was sufficiently close, a filament of matter would escape from the tip of this bulge. Within the filament protoplanetary condensations would form and then would be attracted by the retreating star to give orbital motion around the sun. Initially the orbits would have been elliptic and at the first perihelion passage the action of the sun on the protoplanets would have been like that of the massive star on the sun, so giving rise to a family of satellites. A resisting medium would have been formed around the sun by matter drawn out of it, but not going into the formation of planets, and in this resisting medium the planetary orbits would gradually have rounded off.

The important processes of the tidal distortion and disruption of the sun and the break-up of a filament of matter were subjected to detailed theoretical analysis by Jeans. In Fig. 1.4 we see the effect of an external mass on an extended body as the

Extended body   Tide-raising mass

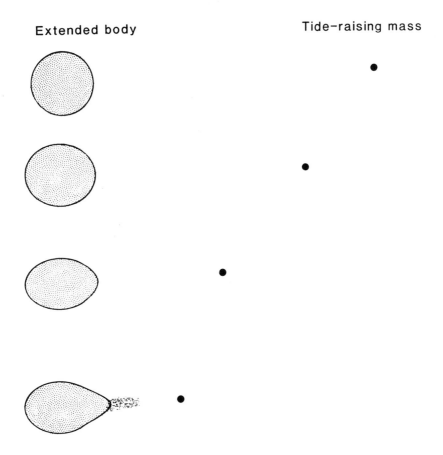

Fig. 1.4 — Behaviour of a body under increasing tidal forces.

distance between them changes. The extended body is at first little different from spherical, then takes up an egg-shape, and eventually develops a pointed end from which material subsequently escapes in the form of a filament. In Fig. 1.5 there are

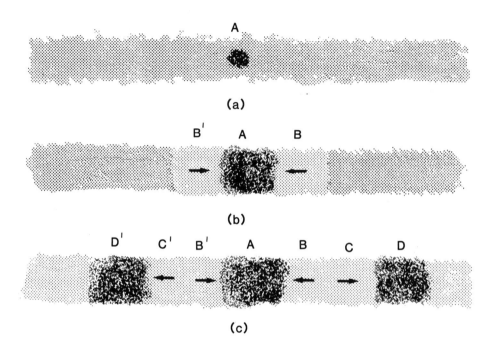

Fig. 1.5 — (a) A filament, uniform except for a small density excess at A. (b) Material at B and B' attracted towards A. (c) Material at C and C' moving away from the lower density regions at B and B' to create higher density region at D and D'.

shown the stages in the break-up of a tidal filament. Firstly we see in Fig. 1.5(a) a uniform filament except that there is a small density excess at the point A. This creates an asymmetry in the gravitational forces on neighbouring material, at points such as B and B', which begins to move towards A as shown in Fig. 1.5(b). The regions outside B and B' now have a deficit of material so that matter at C and C' moves away from B and B' creating higher density regions at D and D' (Fig. 1.5(c)). The disturbance travels outwards from A at the speed of sound in the material and a series of condensations will form in the filament. Jeans showed that the distance between them will be

$$\lambda = \left\{ \frac{\pi}{\gamma G \rho} \right\}^{1/2} c \qquad (1.1)$$

where c is the speed of sound in the gas, ρ its density, γ the usual ratio of specific heats of the gas and G the gravitational constant.

Whether or not a condensation will form a planet depends on its total mass, which, given λ, depends on the line density (mass per unit length) of the filament. For a mass of gaseous material of particular composition, density and temperature there are opposing forces, the lack of balance of which will determine whether the mass will collapse or expand. The force tending to cause collapse is self-gravity while that tending to cause expansion is the temperature-induced pressure of the gas. Jeans found that when the forces balance the critical mass will be

$$M_J = K\frac{c^3}{\{\gamma^3 G^3 \rho\}^{1/2}} \tag{1.2}$$

where K is a constant of order 0.6. Any mass less than M_J will expand and dissipate; any mass greater than M_J will form a condensed body.

The first doubts about this tidal theory were expressed by Jeffreys in 1929. His first concern was that a very massive star would be necessary to give the required tidal effects and such massive stars are very rare. However, a more important point concerned the axial rotations of the sun and the planets. Jeffreys noted that the mean densities of the sun and Jupiter were similar and deduced that material drawn from the sun to produce Jupiter would have undergone about the same degree of condensation as the sun itself. He then quoted a theorem which showed that, in such a circumstance, the rotational periods of Jupiter and the sun should be similar rather than being in the ratio 1 : 60 which is actually observed. There are some weak links in this argument, in particular in assigning the mean density to the early sun's outer region, from which the Jupiter-forming material would have come, and also in ignoring possible rotational effects due to the first perihelion passage of a proto-Jupiter.

To overcome the above difficulty and give a mechanism for imparting rotation to planetary material, Jeffreys suggested that the tidal interaction should be replaced by a grazing collision between the two bodies — a return to Buffon's suggestion. The shear in the boundary layer of the sun would result in rapid rotation of its material and of any planets formed from it.

Another dynamical objection to planets being produced from solar material, either tidally or collisionally ejected from the sun, was raised by Russell in 1935. In Fig. 1.6 there is shown an elliptical orbit with various orbital characteristics shown in terms of the semi-major axis, a, and the eccentricity, e. The semi-latus rectum

$$p = a(1 - e^2) \tag{1.3}$$

is a measure of the intrinsic angular momentum (angular momentum per unit mass) of the orbiting material, which is given by

$$J = (GM_\odot p)^{1/2} \tag{1.4}$$

where $M_\odot$ is the mass of the sun.

The perihelion distance is given by

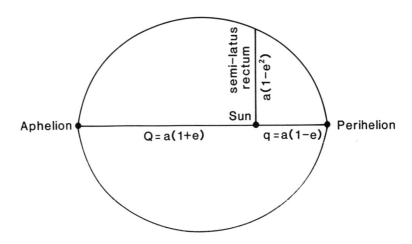

Fig. 1.6 — The characteristics of an elliptical orbit.

$$q = a(1-e) = p/(1+e) \tag{1.5}$$

and it follows that for an elliptic orbit, with $e < 1$, the semi-latus rectum cannot be greater than twice the perihelion distance. For matter coming from the surface of the sun the perihelion distance must be less than a solar radius. This puts a very tight constraint on the intrinsic angular momentum of escaping material and even the most favourable modifications of the orbit of the escaping material, for example by the attraction of the retreating massive star, make little difference to the conclusion. There is in fact another problem that unless the orbit is appreciably changed it will intersect the surface of the sun and so the matter will be reabsorbed. These conclusions were confirmed by computer-based analysis by Lyttleton in 1960.

The basis of Russell's objection would be a strong one if it was taken that the tidal interaction occurred with the sun in a state similar to that at present, since then the angular momentum imparted to escaping material would be insufficient to explain Mercury, let alone the outer planets. However, it should be said that, by 1919, Jeans was assuming that the tidally disrupted sun had a radius of about that of Neptune's orbit (about 4.5×10^{12} m). This would certainly have negated the angular momentum objection, although it introduced other problems such as how newly formed protoplanets, moving towards perihelion, would have interacted with the collapsing sun.

A further difficulty in two-body interaction theories was pointed out by Spitzer in 1939. If the sun was much in its present condition when the stellar interaction took place then, to extract the amount of material to form, say, Jupiter, it would be necessary to draw material from a depth which would result in the material having about the average density of the sun and a temperature of 10^6 K. Testing these conditions in equation (1.2) it would be found that M_J is about 15 Jupiter masses. This is then the minimum mass which could condense to form a planet and no direct condensation of the planets we actually observe seems possible on the basis of this argument. In fact this argument has been challenged; it has been asserted that the initial expansion of the hot solar material can lead to cooling and liquefaction but, nevertheless, it is now generally believed that planets will not form directly from hot solar material. Another observation, which supports the idea that material at high stellar temperatures has not been the raw material for the planets, comes from the distribution of the light elements lithium, beryllium and boron. These substances are rare in the sun, since they are consumed by nuclear reactions at solar temperatures, but they are comparatively abundant in the earth's crust. The conclusion seems clear: that the earth at any rate has never been at solar temperatures.

Actually Spitzer's argument against the tidal theory is invalid for the extended and cool sun model. The sun's mass contained within the radius of Neptune's orbit would have a density of approximately $5 \times 10^{-9}\,\mathrm{kg\ m^{-3}}$ and, with a temperature of say 20 K, the Jeans critical mass would be of order 10^{27} kg, or one-half of the mass of Jupiter.

Despite the bright beginning and enthusiastic reception of the original tidal theory it gradually lost favour. While many of the criticisms referred to above did not apply to Jean's extended and cool solar model, the difficulty of the interaction of newly formed planets and collapsing sun remains. Jeans himself was aware of the problems and wrote 'The theory is beset with difficulties and in some respects appears to be definitely unsatisfactory'. However, it did provide the stimulus for Jeans to analyse many of the situations relating to tidal disruption and this analysis was not invalidated by the downfall of the tidal theory. Later we shall see that use has been made of Jeans' analyses in a more modern context.

2

The structure of the solar system

2.1 INTRODUCTORY REMARKS

In order adequately to discuss the earliest ideas about the origin of the solar system it has been necessary to refer to only the very coarse structure of the system. It is remarkable that, even at this level, the two main theories so far considered were quite inadequate. Neither the spontaneous evolution of a nebula nor Jeans' simple two-body interaction would produce the desired result.

When tackling a new problem one does not know *a priori* whether a plausible solution will be very difficult to find or, conversely, whether almost any model investigated will apparently fit the facts. For the problem of the origin of the solar system it seems that the former situation prevails but, at both these possible extremes, it becomes advantageous to look at theories in relation to the finer details of the system. The fine details of the system, and their interrelationships, may suggest possible models where models are hard to find. Alternatively, where all models seem to explain the first-order description of the system, the details may provide a set of constraints against which the plausibility of models may be tested.

2.2 PLANETARY ORBITS

One of the most striking manifestations of order in the solar system is found in the orbits of the constituent bodies, the characteristics of which are shown in Table 2.1. The way in which the orbital radii agree with Bode's law out to Uranus has already been mentioned, but the law breaks down for Neptune and Pluto.

The two extreme members of the system depart most strongly from circular orbits and from coplanarity with the remainder of the system. Pluto in particular has an orbit with perihelion distance less than that of Neptune. In projection on the plane of the ecliptic (the plane of the earth's orbit) the orbits of these two planets would cross but because of the special relationship of the two orbits the planets never come together closer than 18 AU. Interestingly, for Pluto, the closest approaching planet is Uranus which can come as close as 11 AU.

Table 2.1 — The orbital characteristics of the planets

Planet	Mean distance (AU)	Orbital eccentricity	Orbital inclination
Mercury	0.387	0.2056	7° 0'
Venus	0.723	0.0068	3°24'
Earth	1.000	0.017	
Mars	1.524	0.093	1°51'
Asteroids	2.7	(various — see Table 2.4)	
Jupiter	5.203	0.048	1°18'
Saturn	9.539	0.056	2°29'
Uranus	19.19	0.047	0°46'
Neptune	30.07	0.0086	1°47'
Pluto	39.46	0.249	17°19'

Although the sun contains only a very small part of the angular momentum of the system, it is noteworthy that its spin axis is inclined at 7° to the vector representing the resultant orbital angular momentum of the rest of the system.

2.3 PLANETS AND THEIR STRUCTURES

The basic characteristics of the planets are listed in Table 2.2. With the exception of

Table 2.2 — Characteristics of planetary bodies

Planet	Mass (Earth units)	Diameter (km)	Density (10^3 kg m^{-3})
Mercury	0.0552	4 800	5.5
Venus	0.815	12 100	5.25
Earth	1.000	12 756	5.52
Mars	0.1075	6 790	3.94
Jupiter	318.3	142 200	1.33
Saturn	95.2	119 300	0.71
Uranus	14.7	47 100	1.71
Neptune	17.3	48 400	1.77
Pluto	0.0025	1.445	~1.2

Mass of the earth, $M_\oplus = 5.974 \times 10^{24}$ kg.

Pluto they are clearly divided into two types, separated by the asteroid belt and distinguished by mass and density. This point is illustrated in Fig. 2.1 which depicts

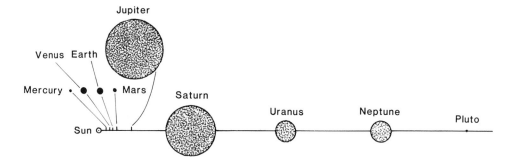

Fig. 2.1 — The relative orbital radii and sizes of the planets. Planets are represented at 10 000 times their natural linear dimensions relative to the depicted orbital radii.

the different sizes of the planets as well as their orbital radii. The minor planets, of which the earth is the most massive member, are dense rocky bodies and almost certainly have metallic cores consisting mainly of iron (Fe) with a small percentage of nickel (Ni). The interpretation of their different densities is in terms of the relative size of the core to that of the whole body. Another common characteristic of the inner planets is that they all show signs of bombardment damage in the form of craters and large depressions. Only a fragmentary picture of the surface of Venus is available and the surface of the earth has been much disturbed by tectonic activity, but the evidence for an intense bombardment of bodies in the inner solar system is quite unequivocal.

There is more uncertainty about the structure of the major planets but it seems likely that they too have metallic and rocky cores of unknown size with an overlay of icy materials, e.g. water and ammonia, together with thick hydrogen and helium atmospheres. Different relative proportions of these components, taken in conjunction with the variation of gravitational compression effects due to their different masses, explain the range of densities of the major planets. There is a school of thought which takes the core of Jupiter to be mainly in the form of solid metallic hydrogen, which ought to exist at very high pressures, and quite convincing models have been built on this basis.

Estimates of the mass of Pluto have steadily declined since the planet was first discovered in 1930. Prior to its discovery it was postulated that a ninth planet should exist and its mass was estimated as six times that of the earth ($6M_\oplus$) to explain small departures of the motion of Uranus from that predicted. By 1978 this estimate had been reduced in several stages to $0.08M_\oplus$ but the discovery of a satellite of Pluto in 1979 gave the current estimate in Table 2.2 of $0.0025M_\oplus$. Since this is only one-fifth of the mass of the moon it seems reasonable to suppose that Pluto is not a conventional planet and that its origin might be ascribed to some special event rather than to the process, or processes, which produced the normal planets.

2.4   PLANETARY SPINS AND SATELLITES

An examination of Table 2.3, which shows the spin periods of the planets and details of their satellite families, suggests that there may be a relationship between these two features. The three giant planets Jupiter, Saturn and Uranus, which rotate most rapidly (assuming that the period of Neptune is at the upper end of the given range), are those with the greatest numbers of satellites. They are also those which possess *regular* satellites; that is to say ones which are in almost circular orbits in the planet's equatorial plane. Such a description may be applied to the satellites of Jupiter out to Callisto, of Saturn out to Hyperion and to all the five main satellites of Uranus. In addition these three planets have ring systems: thin sheets of small bodies orbiting in the equatorial plane. The rings of Saturn are most easily observed and have been most widely studied. They contain annular zones of different appearance with gaps between them (Plate 1); the gaps correspond to distances at which the orbital periods around Saturn are some simple fraction of the periods of some of its inner satellites and are the result of resonance perturbation.

Of the non-regular satellites the moon is an outstanding example, mainly by being so massive in relation to its primary body. Like other rocky bodies in the terrestrial region of the solar system it shows indications of heavy bombardment, these being very little altered by the passage of time. From the point of view of the cosmogonist, the moon, the only other large body of the solar system whose material we can examine at first hand, is of outstanding importance and to explain its features and its relationship to the earth must be one of his primary objectives.

The two small satellites of Mars, Phobos and Deimos, which move in direct orbits very close to their primary body, have been well studied by photography from spacecraft at a short distance. They are of irregular shape and heavily cratered; a Viking picture of Phobos is shown in Plate 2. Their masses have been estimated by the gravitational perturbation they produce on nearby spacecraft and their volumes by photographing them in various aspects. It is concluded that they are rocky bodies of surprisingly low density — perhaps of material similar to carbonaceous chondrites (see section 13.2).

The outer eight satellites of Jupiter shown in Table 2.3 are arranged in two groups of four: an inner group in direct orbits at a mean distance of about 11.5 million km and an outer group in retrograde orbits at approximately twice that distance.

The outermost satellite of Saturn, Phoebe, is also retrograde and is nearly four times further from the planet than the next innermost, Iapetus. This last satellite has the property that its brightness varies greatly as it orbits Saturn, which is explained by the fact that it has one hemisphere with a much higher albedo than the other. This seems to be due to some dark material covering the hemisphere centred on its direction of motion, although the origin of the material is unknown. The dominant member of Saturn's family, the natural satellite Titan, is also unusual in being the only satellite on which a substantial atmosphere has been detected.

Uranus shares with Venus (and Pluto) the distinction of having a retrograde spin. In the case of Venus the spin is slow and the spin axis is almost perpendicular to the ecliptic but Uranus rotates rapidly with its spin axis almost in the plane of the ecliptic. Nevertheless the named family of five satellites listed in Table 2.3 is regular and moves strictly in the equatorial plane of the planet as does the material of the ring system.

The giant planet with the least endowment of known satellites, at least in terms of number, is Neptune. Triton is a large and massive satellite moving in an orbit which is close to the planet and almost circular, but retrograde and well-tilted with respect to

Table 2.3 — The spin characteristics of planets and details of satellites. Planets with retrograde spins, and irregular satellites with retrograde orbits, are labelled (R)

Planet/satellite	Spin period	Inclination of spin axis to ecliptic or of orbit to planet equator	Orbital radius (10^3 km)	Orbital eccentricity	Mass (10^{22} kg)	Average diameter (km)	Density (10^3 kg m^{-3})
MERCURY	58.7 d	0°					
VENUS	243.0 d	178° (R)					
EARTH	24 h	23°27'					
Moon		23.4°	385	0.055	7.35	3476	3.34
MARS	24 h 37 m	25°12'					
Phobos			9.4	0.021		23	2
Deimos			23.5	0.003		13	2
JUPITER	9 h 55 m	3° 7'					
Adrastrea			127			40?	
Metis			129			20?	
Amalthea	0.4		180	0.003		200	
Thebe			222			80?	
Io		0.0°	422	0.000	8.93	3652	3.5
Europa		0.5°	671	0.000	4.88	3126	3.0
Ganymede		0.2	1070	0.001	14.97	5276	1.9
Callisto		0.2°	1885	0.01	10.68	4820	1.8
Group of four		25 to29°	11110 to 11740	0.130 to 0.207		10 to 170	
Group of four		147 to 164° (R)	20700 to 23700	0.169 to 0.378		20 to 40	

SATURN	10 h 14 m	26°45'					
Atlas		0.3°	138	0.002		60	
Shepherd		0.0°	139	0.003		110	
Shepherd		0.05°	142	0.004		90	
Epimetheus		0.1°	151	0.007		200	
Janus		0.3°	151	0.009		60	
Mimas		1.5°	186	0.020		390	1.2
Enceladus		0.0°	238	0.005		510	1.1
Tethys		1.9°	295	0.000		1050	1.0
Telesto			295			30	
Calypso			295			25	
Dione		0.0°	378	0.002		1120	1.4
Dione B			378			30	
Rhea		0.4°	527	0.001		1530	1.3
Titan		0.3°	1222	0.029	14.22	5150	1.9
Hyperion		0.4°	1483	0.104		280	1.9
Iapetus		14.7°	3560	0.028		1440	1.2
Phoebe		159° (R)	12950	0.163		200	200
URANUS	17.2 h	97°53' (R)					
Ten small satellites			31 to 53			10 to 100	
Miranda		0.0°	131	0.000		480	
Ariel		0.0°	192	0.003		1160	
Umbriel		0.0°	268	0.003		1180	
Titania		0.0°	439	0.002		1580	
Oberon		0.0°	587	0.007		1540	
NEPTUNE	15–24 h?	28°48'					
Triton		159.9°(R)	356	0.000		3700	
Nereid		27.7°	5570	0.749		300	
PLUTO	6.39 d	118° (R)					
Charon		0.0°					

the planetary equator. Its companion satellite, Nereid, is also remarkable in having the largest orbital eccentricity (=0.75) of any known satellite.

The final satellite is that of Pluto, aptly named Charon, to which reference has already been made. It has been detected only as a distortion of the image of Pluto, but its period is quite well-known even if its orbital radius can only be crudely estimated.

From the widespread distribution of satellites within the solar system it might be deduced that they would be a common feature of planetary systems in general — assuming that other systems exist. However, it is also clear from the varied relationships between satellites and their planets that a single mechanism of origin may not suffice to explain all of them.

2.5   SMALLER BODIES OF THE SOLAR SYSTEM

We have previously seen that the minor and major planets are divided by the asteroid belt in the region centred on 2.8 AU from the sun which would have accommodated a Bode's-law-predicted planet. Indeed when in 1801 Giuseppe Piazzi discovered Ceres, the largest of the asteroids, it was believed that the pattern revealed by Bode's law was complete, but shortly afterwards, and in quick succession, Pallas, Juno and Vesta were discovered. Now thousands of these small bodies are known and the characteristics of some of them are shown in Table 2.4.

Table 2.4 — The charactistics of some important asteroids

Asteroid	Radius (km)	Mean orbital radius (AU)	Eccentricity	Inclination
Ceres	350	2.8	0.079	10.6°
Pallas	230	2.8	0.235	34.8°
Juno	110	2.7	0.256	13.0°
Vesta	190	2.4	0.088	7.1°
Hygiea	160	3.2	0.099	3.8°
Eros	7	1.5	0.223	10.8°
Icarus	0.7	1.1	0.827	23.0°
Apollo	0.5	1.5	0.566	6.4°
Hermes	0.3	1.3	0.475	4.7°

The eccentricities and inclinations of the asteroids tend to be much higher than those of the planets but they are all in direct orbits. A distribution of the orbital radii shows that the majority have a mean value in the range 2.2–3.2 AU with an average close to 2.7 AU. There are, however, distinct breaks, known as Kirkwood gaps, in the distribution of orbital periods as they correspond to simple fractions — $\frac{1}{2}, \frac{3}{7}, \frac{2}{5}, \frac{1}{3}$, etc. — of the orbital period of Jupiter. We have already mentioned similar gaps in Saturn's rings due to resonance perturbation by satellites.

The other kind of smaller object visible from the earth is the comet. Comets are bodies with highly eccentric orbits which occasionally bring them into the inner reaches of the solar system. Under solar irradiation some of their material vaporizes and they develop diffuse luminous heads and characteristic tails which, under the influence of the solar wind, point away from the sun. Comets tend to divide themselves into two categories according to the period of their orbits. The short-period comets reappear regularly and records of Halley's comet go back to 240 B.C. On the other hand some comets have orbits which are so elongated that it strains the accuracy of observation to deduce whether their orbits are elliptical or hyperbolic. The general view is that they are all, or nearly all, elliptical with periods upwards of a million years. On this basis, at aphelion they would be several tens of thousands of astronomical units from the sun.

Since we can only see comets when they come close to the sun it is reasonable to suppose that what we see is a sample of a much larger population, most of which will never be seen. The general belief is that there exists, mostly at great distances from the sun, a reservoir of about 10^{11} comets some of which are occasionally perturbed by nearby stars into visible orbits. We shall examine the implications of this hypothesis later (section 13.1).

From the spectra of the light from comets there may be detected a variety of compounds and radicals involving the elements carbon, nitrogen, oxygen and hydrogen, from which it may be deduced that comets have an icy composition, or at least partly so. At each passage past the sun a comet must lose a proportion of its material and hence its lifetime as a visible object must be limited. In their passages through the solar system comets are often appreciably perturbed by the planets, particularly the major ones, and long-period comets may occasionally be perturbed into short-period orbits.

Another class of small solar-system objects is the meteors, which are seen as streaks of light when they enter the earth's atmosphere. These small bodies are often organized in the form of a meteoroid stream: a myriad of small objects in the same elliptical orbit around the sun with the material spread out around the complete orbit. The intersection of this orbit with that of the earth will give rise to a periodic display of meteor showers. For example, every year at about mid-August the earth in its orbit crosses one such meteoroid stream and we see the sometimes spectacular Perseid shower. There is strong evidence that meteoroid bodies may be the debris from comets, since the orbits of some showers coincide with those of short-period comets. Nevertheless there is also a great deal of uncoordinated debris in the solar system and the zodiacal light, a band of light in the sky, best seen when the sun is not too far below the horizon, is caused by sunlight scattered by small bodies occupying the region between the earth and the sun.

By far the most interesting and informative small bodies in the solar system are meteorites: meteor-like objects which do not completely vaporize but which strike the earth and may be recovered. A first-order rather crude classification of meteorites is in terms of being irons, stony-irons or stones, which makes it tempting to think of them as components of a layered planet like the earth. However, detailed investigation of their physical structure and chemical and isotopic compositions reveals a wealth of information about conditions which must have prevailed in the early solar system. For example, iron meteorites, which also contain a small percentage of nickel, show characteristic patterns, called Widmanstätten figures (Plate 3) in polished cross-sections. This is due to the migration of nickel atoms between two phases of nickel–iron in a solid lattice of the hot metal, and from the scale of the pattern can be deduced the cooling rate of the specimen. Again, most stony meteorites are so-called chondrites which contain glassy spherical or near-spherical inclusions (Plate 4), another indication that in the early solar system temperatures were at least high enough to melt silicate materials. We shall return to these matters in some detail in section 13.2.

3

Later ideas

3.1 THE NEBULA WITH TURBULENCE

With the demise of the Jeans tidal theory a position had been reached where two fairly simple models had been tried and had been found wanting. As a natural consequence of this situation, theorists turned their attention to more complicated models and, indeed, it became the unspoken but unchallenged assumption of cosmogonists that only in some theoretically complicated and difficult model would the truth be found.

Even during the heyday of the tidal theory there were still those who steadfastly adhered to the nebula concept and looked for ways in which its difficulties could be resolved. In the early 1930s some quite elaborate work was done by Berlage on the effects of viscosity in the evolution of a protoplanetary disk left behind by a nebula. This work concentrated on the disk and did not concern itself with the dominant difficulty of the nebula theory — that it would give a rapidly rotating sun, even to the extent that the core of the nebula would have so much angular momentum that no centrally condensed star would form at all.

In 1944 von Weizsäcker put forward a model of a protoplanetary disk which involved the setting up of a pattern of turbulence-induced eddies as shown in Fig. 3.1. The motion around each vortex is in a clockwise sense, but the whole system rotates around the central condensation in an anticlockwise direction. The combined effect of these two rotations, if they are suitably geared, is that an individual portion of matter, say that represented by the point P, moves in a Keplerian orbit around the central mass. In such a system there would be very little dissipation of energy within vortices except that, at points at the boundary between vortices, there would be a meeting of material with a high relative velocity. In von Weizsäcker's picture, within such eddies, roller-bearing eddies are formed and it is in these small eddies, where matter is interacting a great deal, that condensations would form. These condensations would be formed in a series of rings and, when all the condensations in each individual ring had coalesced, a series of planets would have been formed. With five vortices to a ring it was shown that the progression of orbital radii of the planets would have given something similar to Bode's law.

Rotation of whole system

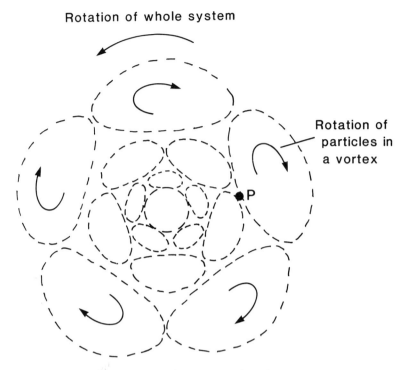

Fig. 3.1 — The von Weizsäcker vortex configuration.

Von Weizsäcker's ideas have been very heavily criticized, especially by Jeffreys in his 1952 Bakerian lecture to the Royal Society. The critics have argued that such a system of vortices could never have become established and that turbulence, which is a phenomenon of disorder, could not be expected to exist in the form of the highly structured and ordered von Weizsäcker model. The more natural end-product for a disk with viscosity-induced turbulence is a quietly rotating system with all parts in circular orbit around the central mass. The vortex system is one of much higher energy and it is inconceivable that it could become established or that it would be stable.

Another problem with planet formation in a disk is that the disk material needs to have a certain minimum density before a condensation can form in the presence of the central body. We can see this by a simple analysis based on the situation shown in Fig. 3.2 where there is considered the condensation of the small spherical mass of density ρ and radius r in the presence of the mass M. The gravitational force of the sphere on a unit mass (i.e. induced acceleration) at a point P directed towards O, F_{PO}, must exceed the gravitational force due to M pulling the mass at P *away from* O, F_{PM}. This latter is given by

$$F_{PM} = \text{Force of } M \text{ on a unit mass at P} - \text{Force of } M \text{ on unit mass at O}$$
and the condition for the spherical mass to condense becomes

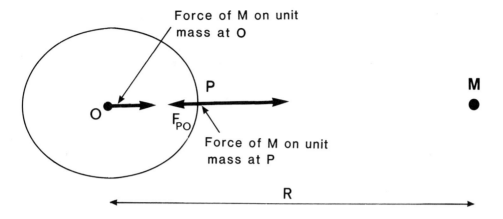

Fig. 3.2 — The system of forces giving equilibrium at the surface of a fluid sphere subjected to the tidal influence of a nearby body.

$$\frac{4\pi r^3 G \rho}{3} \times \frac{1}{r^2} \geqslant \frac{GM}{(R-r)^2} - \frac{GM}{R^2} \tag{3.1}$$

which, for $R \gg r$, leads to the condition

$$\rho \geqslant 3M/(2\pi R^3) \tag{3.2}$$

This gives the minimum density of material which can condense at some distance R from the mass M. The value found is 0.5 times the density of the central mass, M, occupying a sphere of radius R. We may note that in Jeans' criticism of Roche's attempt to rescue the Laplace nebula theory (section 1.2), in a slightly different context, the corresponding factor was 0.361. Equally, if the density is fixed, then we can find the minimum distance, R_1, at which a condensation can form. By rearranging equation (3.2)

$$R_1 = \{3M/(2\pi\rho)\}^{1/3} \tag{3.3}$$

which is called the *Roche limit* for the particular central mass and density of the body in question. When the small spherical body is rotating or where r is comparable in magnitude to R, other forms may be found for the Roche limit differing from equation (3.3) by a numerical factor (usually less than 2).

The limiting density given by equation (3.2) implies a certain minimum mass for a disk if planets are to form in it. It can be shown that if the disk has a fixed thickness, t, and the central mass is the sun then the mass of the disk is, approximately,

$$M_D = 3M_\odot t/R_1 \tag{3.4}$$

where R_1 is the minimum distance at which planets are to be formed. Making the extreme assumption that the disk is very thin with $t = 0.1$ AU and that R_1 is the radius of Jupiter's orbit, then M_D is about $0.06 M_\odot$. This is clearly a gross underestimate yet it is about 40 times the mass of all the planets combined and shows that somehow or other most of the mass must be disposed of. Jeffreys argued that the energy required to remove it would not be available. However, recent observations have suggested that during the T-Tauri stage of the development of a star, if it occurs, a powerful solar wind exists that might well be capable of sweeping away a great deal of mass in its vicinity. We shall return to this in section 3.5.

What von Weizsäcker's theory has very little to say about is the angular momentum problem and it provides no mechanism that can lead to a slowly rotating sun. This must be regarded as an essential feature for any plausible theory in which the same nebula produces both the planets and the sun.

3.2   A NEBULA BY CAPTURE

In 1944 it was suggested by O. Y. Schmidt that the sun could have captured an envelope of gas and dust during a passage through an interstellar cloud. Schmidt assumed that a two-body capture process could not take place and that there was a third body, another star, involved in the process. In 1961 Lyttleton put forward a similar idea but in a form which dispensed with the need for a third body. He assumed that a process of line accretion, proposed by Bondi and Hoyle in 1944, would apply. The passage of the sun through a cloud can be envisaged as in Fig. 3.3 where the

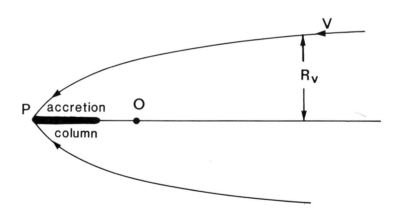

Fig. 3.3 — The Bondi and Hoyle accretion mechanism. Interacting streams destroy the component of velocity perpendicular to the axis, leaving material in an accretion column with less than escape speed.

cloud material moves at a speed V relative to a stationary sun at point O. The cloud material all crosses the axis OP and, if it strongly intereacts along the axis so that the component of velocity perpendicular to that axis is destroyed, then the residual velocity of the material may be less than the escape velocity so that it will be captured. If the stream crossing the axis at P is the furthest one that can be captured and if the stream approaches from a large distance at a distance R_V from the axis then it can be shown that

$$R_V = 2GM_\odot/V^2 \tag{3.5}$$

To allow for effects due to the temperature of the cloud Bondi (1952) suggested a modification of the formula to give a capture radius

$$R_B = 2GM_\odot/(V^2 + c^2) \tag{3.6}$$

where c is the speed of sound in the cloud material. After passing through the cloud the sun would have captured a cylinder of cloud material of this radius.

Lyttleton considered the passage of the sun through a cloud of dimension $L \, (\gg R_B)$ with density ρ and an intrinsic angular velocity ω (Fig. 3.4). The total mass

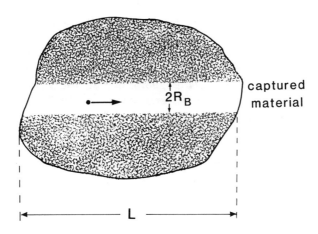

Fig. 3.4 — Passage of the sun through a cloud, and the capture of a cylindrical portion of it by the Bondi and Hoyle mechanism.

of material captured would have been

$$m = \pi R_B^2 L \tag{3.7}$$

and, if it is treated as being derived from a long thin rod rotating about an axis perpendicular to its length, then its angular momentum would have been

$$H = mL^2\omega/3 \tag{3.8}$$

Lyttleton suggested values for the various parameters from general knowledge of the properties of interstellar material. He assumed that the cloud was in thermal equilibrium with the background galactic radiation at 3.18 K and that it consisted mainly of molecular hydrogen. This gave $c^2 = 3.86 \times 10^4 \, \text{m}^2 \, \text{s}^{-2}$. He took the density of the cloud material to be $10^{-20} \, \text{kg m}^{-3}$ and its intrinsic angular velocity to be the same as that of the galaxy, $10^{-15} \, \text{s}^{-1}$. Next Lyttleton took m and H as those required to explain the planets, 3×10^{27} kg and 4×10^{43} kg m^2 s^{-1} respectively, and deduced what the other quantities had to be. This gave L about 10^{16} m, R_B about 5×10^{15} m and $V = 0.2$ km s^{-1}.

The first thing that is noticed is that the condition that $L \gg R_B$ does not hold, but this is not of great importance as far as the acceptability of the theory is concerned. More difficult to accept are the very low relative speed of the sun and the cloud, and the temperature of 3.18 K. The proper speed of the sun in the galaxy is quite typical for field stars of its type, 20 km s^{-1}, and it is not very likely that it would have such a low speed relative to a cloud that it happened to meet. Indeed, since the sun and the cloud have masses and would accelerate towards each other such a speed of contact is impossible! If the relative speed at a great distance was very tiny then the minimum speed at which the sun would meet the cloud is about 0.3 km s^{-1}; any lesser speed is impossible and to be close to that limiting speed is very unlikely. The temperature is also much lower than observation would suggest and a temperature in the range 20–100 K would be much more acceptable. Both these deviations from Lyttleton's assumptions decrease the value of R_B and make it impossible to satisfy equations (3.7) and (3.8).

In 1973 Aust and Woolfson critically analysed Lyttleton's capture mechanism and pointed out that considerable tidal distortion would take place when the sun approached a diffuse cloud. A tongue of matter would be drawn out of the cloud towards the sun and this could be captured, but not by the mechanism suggested by Lyttleton. This point is further discussed in Chapter 6.

Aust and Woolfson did not just criticize Lyttleton's choice of parameters, but they went on to find alternative parameters which would be acceptable. Lyttleton's assumption that all the captured material went into planet formation is not very reasonable since one might expect some material either to be lost from the system or to be absorbed by the sun. As an example of some results given by Aust and Woolfson: if the cloud density was 10^{-19} kg m^{-3}, its temperature 20 K and the sun reached the boundary of the cloud with free-fall speed then $m = 5.6 \times 10^{27}$ kg and $H = 1.9 \times 10^{44}$ kg m^2 s^{-1}, which values are reasonable if allowance is made for some loss of material.

By assuming the separate formation of the sun and planets the angular-momentum problem for the planetary material is essentially solved and any difficulties about the slow rotation of the sun are simply sidestepped. However, it still leaves what turns out to be the very intractable problem of forming planets starting with very diffuse material. Since it is a feature of a number of theories that planets must be formed in this way, we shall now turn our attention to this matter.

3.3 PLANET FORMATION FROM DIFFUSE MATERIAL

A common feature of all nebula theories, no matter how the nebula comes about, is that planets must form from very diffuse material. The picture we have is of a rotating disk-like envelope around the central condensation consisting of gas and of solid particles of ice, silicates and metal. In the terrestrial region only silicate and metal particles would have condensed; for Jupiter and Saturn all types of original material would be present, although with a slight enhancement of the solid component; while for Uranus and Neptune the enhancement of solid material would have been much greater.

There are good reasons to suppose that the original nebula could not have had a mass more than a few hundredths of the mass of the sun. One mechanism by which material could subsequently be lost is by the action of a solar wind during a possible T-Tauri stage of the sun's evolution. Observations of such stars give evidence of matter streaming away from them at speeds of, typically, 100 km s^{-1} and some theorists claim that the mass loss could be one-tenth of that of the sun in a period of 10^7 years. On the other hand there are some doubts about this conclusion and Herbig (1978) says that there may not be any mass loss at all. A speed of 100 km s^{-1} is below the escape speed from the surface of the sun, but if we interpret it as the characteristic speed of outflow at, say, 1 AU then this is twice the escape velocity and, in principle, it could have swept away up to three times its own mass of a nebula. In fact it could hardly do this; a disk nebula of the kind we are considering would be very thin, only about 0.1 AU thick for gas in the terrestrial region, so that only little of the T-Tauri wind could have interacted with it. Again, the process of driving out material would only work for gas or very fine solid particles of radius less than $100 \mu m$ since it is necessary for the wind force to exceed that of gravity for escape to take place.

For larger particles loss from the nebula might take place via the Poynting–Robertson effect. This is due to an interaction of radiation with the material, which removes angular momentum and so causes small particles to spiral in towards the sun. We imagine the solar radiation as a radial stream of photons which are absorbed by the solid particles (Fig. 3.5(a)). However, the photons which correspond to the energy re-radiated by the particle are emitted isotropically *with respect to the particle* and hence they have a component of velocity, corresponding to that of the particle itself, transverse to the radius vector (Fig. 3.5(b)). The angular momentum of the radiation corresponding to this transverse motion comes from the particle which therefore spirals in towards the sun. A straightforward calculation shows that the time for a particle to move from a distance R_o to a distance R from the sun is given by

$$t = 4\pi c^2 a\rho (R_o^2 - R^2)/(3L_\odot) \tag{3.9}$$

where a and ρ are the radius and density of the particle, c the speed of light and $L_\odot$ is the luminosity of the sun ($3.9 \times 10^{26} \text{ W}$). Thus for a particle of radius 10 mm and density $3 \times 10^3 \text{ kg m}^{-3}$ in the vicinity of the earth, the time to be absorbed by the sun is 2×10^7 years and it is clear that much larger particles could have been absorbed in the available timescale. While this may be an attractive scenario for the removal of nebula material, not much material could actually have been removed in this way. After spiralling inwards, at the moment of its absorption by the sun, the particle would have been in orbit around the solar equator. It is easy to show that one Jupiter mass ($M_\odot/1000$) in orbit around the solar equator would have three times as much angular momentum in orbital motion as is possessed by the present sun in its spin. If an appreciable fraction of a solar mass had been absorbed then the sun would now be rotating much more rapidly than it does.

The above arguments support the assertion that planetary formation in a nebula would need to take place with a total mass of nebula material less than a few hundreths of a solar mass and that models which assume more than this mass must also explain how the extra material was eliminated from the system.

We now turn to the problem of the accretion of the nebula material to form planets. The first stage of the process, about which there seems to be little disagreement, is that solid particles will settle towards the central plane. The

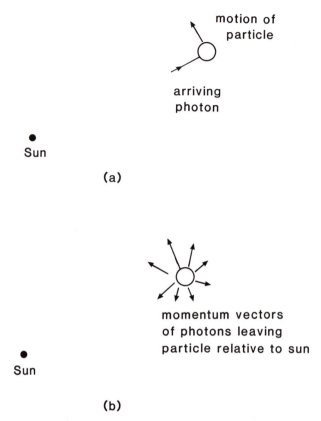

Fig. 3.5 — The Poynting–Robertson effect.

timescale for this is short: of the order of 100 years for particles of radius 10 mm in the vicinity of Jupiter. It is what happens after the creation of this dust carpet which is not so clear.

    One possibility is that the overall disk has a high enough density to satisfy equation (3.2) and hence spontaneously to form condensations. In our previous discussion of this question we considered only the self-gravitational forces within the medium and the dispersive tidal effects of the central body. In 1972 Safronov made a much more detailed study of the case of a thin rotating disk and deduced that the critical density was

$$\rho_c = 6.2 M_\odot/(4\pi\rho^3) \tag{3.10}$$

which is not very different from equation (3.2). However, Safronov's analysis of stability within the disk included many features which the previous analysis had not, including the effect of the thermal energy of the gaseous component and the energy of random motion of solid particles. Thus he found that in a disk of thickness h the volume of a condensation would be of the order $60h^3$. It turns out that any idea of the formation of gaseous condensations is not tenable since it would require the gas to have a temperature less than 1 K. However, it is not inconceivable that solid agglomerates could form. In the vicinity of Jupiter it would require that the average density of solid material would be 2×10^{-6} kg m^{-3} in a disk of thickness 50 000 km. The speed of random motion of the solid particles would need to be less than 0.5 m s^{-1} and then Safronov's result is that the solid bodies formed would have a mass of 2×10^{19} kg with radius of order 100 km. In 1973 Goldreich and Ward disputed this result, claiming that such bodies would not form directly; from thermodynamical considerations they deduced that small planetesimals of dimensions a few hundred metres would be the first stage of planetary formation.

The next stage of the process would have been the development of random motion of the planetesimals which, initially, were in near-circular orbits. Gravitational interactions between planetesimals would have led to an increase in their orbital eccentricities and hence to an increased probability of further interactions. Random motions would tend to grow and eventually collisions between bodies would have become frequent.

There have been many attempts to model the evolution of a swarm of colliding planetesimals. A collision between a pair of planetesimals can lead either to fragmentation and dispersal of one or both of them or to their aggregation. The latter situation will prevail if the approach velocity of the two bodies when they are a large distance apart is appreciably less than the escape velocity from the larger body. In such a case, from the point of view of the larger body, the smaller one will strike it at little more than the escape speed. The smaller body will share its energy with some of the material of the larger one giving rise to many fragments, all with less than the escape speed from the larger body which will subsequently accrete them. The larger the body the greater would have been its accretion rate and this would have led to a runaway growth with one large body tending to dominate within each planetary zone. An impression of the total mechanism for producing a planet, as suggested by Safronov, is given in Fig. 3.6.

Safronov calculated the characteristic timescales for planetary growth. In the terrestrial region he found timescales of 10^7 years but the time estimates increased rapidly in the outer regions of the solar system and was 10^{10} years for Neptune — which is twice the age of the solar system.

It is clear that, in view of the large timescales found for the formation of the outer planets, a satisfactory theoretical model for the accretion of planets from diffuse material is not available at present.

3.4   THE FLOCCULE THEORY

In 1960 McCrea described a theory in which the process of forming planets is directly linked with the formation of stars in a cluster. One of the observational problems with which this theory deals is that of the slow rotation of the sun. The rotational characteristics of stars will be discussed more fully in Chapter 5, but for now we may note that the equatorial speed of solar material is 2 km s^{-1} and that low equatorial speeds are characteristic of stars with masses less than about $1.4 M_\odot$. This indicates that such stars, or indeed any stars, could not be formed by the direct and isolated collapse of galactic material. If such material, with density 10^{-21} kg m^{-3} and intrinsic

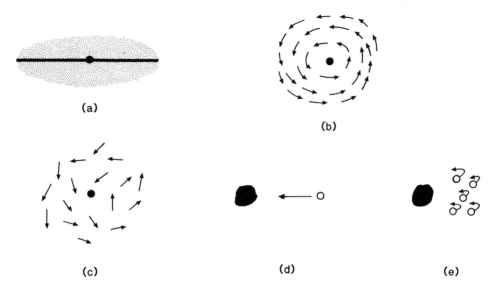

(a)

(b)

(c)   (d)   (e)

Fig. 3.6 — The Safronov model of planetary formation. (a) A dust disk forming in the nebula. (b) Solid condensations take up circular orbits. (c) Perturbations give non-circular, intersecting orbits. (d) An impending collision between two objects with a high mass ratio. (e) The smaller object, plus fragments of the large one, are accreted by the more massive body.

speed of rotation equal to that of the galaxy, 10^{-15} s^{-1}, is compressed to solar density then the equatorial speed would be several times the speed of light!

McCrea's model, presented in a highly developed form in 1978, starts as a dense cloud of interstellar matter which is going to form a stellar cluster. The collapse will be turbulent and the collision of turbulent streams will lead to regions of higher than average density. McCrea models this by taking the cloud to consist of floccules: isolated high-density regions each of mass m, radius s and density ρ_O which move with a mean randomly directed speed V with respect to the centre of mass of the system (Fig. 3.7). This speed is taken as the speed of sound in the cloud material for it is argued that, although the turbulence will be supersonic, it would be difficult to have speeds much higher than this.

Floccules occasionally collide and when they do so they combine. At intervals throughout the cloud, just by chance, a larger than average aggregation will form and will begin gravitationally to attract other floccules, thus forming a protostar. If a star is formed within a region of radius S within which there are N floccules then it can be shown that the expected angular momentum in the region will be

$$H = \tfrac{1}{2}mVSN^{1/2} \tag{3.11}$$

Actually the idea of an isolated region forming a star is of dubious validity since floccules will move from region to region but McCrea assumed that the net effect is equivalent to having isolated regions with floccule-reflecting walls.

A protostar will have a mass greater than the critical Jeans mass (equation (1.2))

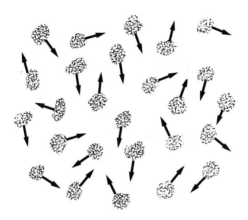

Fig. 3.7 — Floccules in random motion.

and so it will contract and form a smaller target. Any floccule which strikes the protostar must have been on a path whose original line of motion was not too far from the centre of the protostar and hence will tend to contribute little angular momentum to it (Fig. 3.8). An additional factor is that the floccules will strike the protostar from

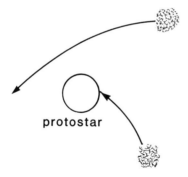

protostar

Fig. 3.8 — A floccule joining a protostar and contributing little angular momentum.

random directions and so will tend to cancel out the contributions to the angular momentum. McCrea showed that if nearly all the floccules in a given star-forming region ended up in the central body then its spin angular momentum would be

$$H_S = \left\{ \frac{8V_O^2 R_S}{27V^2 S} \right\}^{1/2} H \tag{3.12}$$

where R_S is the final radius of the star and V_O is the escape speed at a distance S from the star. Since R_S is so much smaller than S the ratio H_S/H is very small.

Next McCrea argued that the residual floccules, N_1 in number, which did not go into the star, must account for virtually all the angular momentum in the region and these will have collected together in small aggregations to form a system of planets. The masses of the planetary aggregations are expected to be not much greater than M_J, the Jeans critical mass, for the floccule material since, once this mass is reached, the aggregations would have condensed to form a small target for the few floccules which remain.

McCrea's approach in considering numerical aspects of his model was to take some parameters with their usually accepted values, shown as adopted values in Table 3.1, while others were postulated and chosen to give a best fit with observational quantities of the solar system. With the values given in Table 3.1, the

Table 3.1 — The parameters of the floccule theory

Adopted parameters
    Mass within star-forming region $M = 2 \times 10^{30}$ kg
    Temperature of material $\theta = 50$ K
    Mean turbulent speed $V = 1$ km s^{-1}
    Mean molecular weight of material $\mu = 2$
Postulated parameters
    Radius of star-forming region $S = 10^{10}$ km
    Number of floccules in region $N = 10^5$
    Density of floccule material $\rho_O = 10^{-6}$ kg m^{-3}
Derived parameters
    Floccule mass $m = 2 \times 10^{25}$ kg
    Floccule radius $s = 1.7 \times 10^7$ km
    Jeans mass for floccule material $M_J = 9.5 \times 10^{26}$ kg
     $\simeq 48m$

predictions of Table 3.2 were found. Many of the predictions are quite good, especially bearing in mind the approximations of the model, but the outstanding discrepancy concerns the spin angular momentum of the principal planets. If n floccules come together to give a planet then, by a result similar to equation (3.11), the expected spin angular momentum will be

$$H_P = \tfrac{1}{2} m v_P r_P n^{1/2} \tag{3.13}$$

where r_P is the radius of a mass M_J at floccule density, v_P is the escape speed from the surface of such a body and $n = M_J/m$. This yields the value given in Table 3.2.

The other apparent anomaly in Table 3.2 is the statement that there are only six principal planets. This is coupled with the angular momentum disagreement and is to do with the way in which McCea resolves that problem, which will now be explained.

Table 3.2 — Predictions of the floccule theory

Property	Empirical	Predicted
Mass of sun (kg)	1.99×10^{30}	2×10^{30}
Angular momentum of sun (kg m^2 s^{-1})	1.6×10^{41}	7.4×10^{41}
Total mass of planets (kg)	2.7×10^{27}	0.9–20×10^{27}
Radius of system (AU)	39	3–67
Orbital angular momentum of planets (kg m^2 s^{-1})	3.1×10^{43}	3.2×10^{43}
Number of principal planets	6	1–20
Mass of principal planets (kg)	1.9×10^{27}	10^{27}
Principal planet spin angular momentum (kg m^2 s^{-1})	0.42×10^{39}	4.4×10^{39}

As a background to this explanation we need to know the form of development of a rotating and collapsing, initially spherical, mass. The rotational speed of the body increases as it collapses until, eventually, a state of rotational instability is reached. The loss of material then takes place as envisaged by Laplace to give a disk in the equatorial plane. However, the central collapsing core eventually reaches a high density where, for all practical purposes, it behaves like an incompressible material After going through being the shape of a prolate spheroid (Fig. 3.9(a)) it then passes to a general ellipsoidal shape (Fig. 3.9(b)) and next to an unstable pear-shape (Fig. 3.9(c)) which finally breaks up into two unequal parts with a mass ratio greater than 8:1 Fig. 3.9(d)). At this stage the original angular momentum has been converted into relative motion of the two bodies as well as the spins of each of them.

McCrea now follows a suggestion of Lyttleton (1960) that the planets Jupiter, Saturn, Uranus and Neptune are the larger portions of such fission processes, with the smaller portions having been ejected from the solar system. Such a suggestion is quite feasible since the smaller portions would have a high velocity with respect to the centre of mass of the system. The larger portion would have retained most of the mass of the original body but only a small fraction of the angular momentum which was mainly contained in the relative motion of the two parts. From an analysis of the break-up model, McCrea has explained quite well the observed rotational periods of Jupiter and Saturn.

For the major planets it was assumed that the gas and dust components of the original protoplanets stayed well·mixed in the period up to fission. The other possibility is that there could have been sedimentation of the dust to form a core before the body as a whole had substantially collapsed. McCrea described a process where it was the dusty core itself which underwent the process of fission. If core fission took place in the terrestrial region of the solar system then it is possible that both portions would have been retained since neither would have acquired escape speed, which is higher in the inner part of the solar system. It was suggested that the planet pairs Earth–Mars and Venus–Mercury arose in this way, and McCrea has

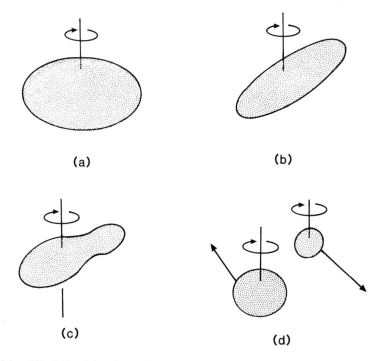

Fig. 3.9 — The fissional break-up of a rotating and collapsing pseudo-incompressible fluid sphere. (a) A prolate (MacLaurin) spheroid. (b) A general (Jacobi) ellipsoid. (c) An unstable pear-shaped configuration. (d) Two bodies produced by fission.

pointed out that these combinations have very similar mean densities and metal fractions and also mass ratios greater than the 8:1 that theory requires. The numerical predictions for the terrestrial planets are not as satisfactory as for the major planets, since rotation periods of about four hours are predicted, but it has been pointed out that these might have been modified by tidal action, perhaps by the sun, after the planets were formed.

McCrea, again following Lyttleton, ascribed the origin of satellite systems to the formation of droplets between the separating components of the fission process (Fig. 3.10). Lyttleton showed that such droplets could escape completely from the

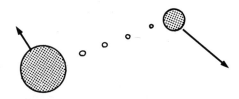

Fig. 3.10 — Droplets between fission-produced bodies giving a satellite family.

two portions or could be retained as satellites by the larger one, but not by the smaller.

The floccule theory has the very satisfactory feature that it deals with many aspects of the solar system in a coordinated way and that detailed calculations can be made so that theory and observation may be compared. Many of the processes are quite convincing, such as the formation of a slowly rotating sun and the origin of satellite systems. Nevertheless when the theory is examined in depth it appears that there are many severe problems, not least with the floccules themselves. While floccules are only an attempt to model turbulence, they should have properties which are related to those of a real turbulent system; the predictions appearing in Table 3.2 are sensitively dependent on the postulated parameters.

Many accounts have been given of the collision of turbulent streams of gas, and for the head-on collision of two streams, each moving with the speed of sound, the compression factor is about 2.5. If compressed regions subsequently interact then further compression will result, but it is not really possible to achieve compressions of three orders of magnitude, which is what floccule formation requires. If the conditions are not isothermal then the compression is even less than that quoted above.

The second problem concerns the lifetime of individual floccules or even small collections of floccules. Any assemblage of radius r and mass much less than M_J will expand and dissipate in a time of order r/c, where c is the speed of sound in its material. For McCrea's floccules this time is about one year. However, the average time between floccule collisions, which is necessary to build up aggregations, is about 30 years; it seems very unlikely that a coherent mass M_J, equivalent to a few tens of floccules, could actually be built up.

A third doubt concerns the assumption that the regions of radius S in which individual stars form can be regarded as isolated. Floccules do pass from region to region and so the regions are highly coupled. As an illustration we see in Fig. 3.11

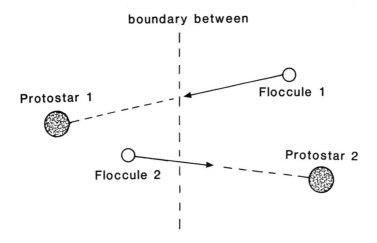

Fig. 3.11 — Coupling between neighbouring star-forming regions.

two neighbouring regions, each with its developing protostar, and with a floccule from each region following a path leading to incorporation in the protostar of the other region. Certainly they will contribute little to the angular momentum of the protostars they join, but equally it is certain that angular momentum has been lost from the regions in which they originated and is not available to be shared by the residual floccules supposed to form planets. The angular momentum, which is conserved, has now gone into the relative motion of the two stars.

Finally it was shown by Aust and Woolfson in 1971 that, statistically speaking, the floccule theory is not likely to give an initially planar system and might indeed give one or two retrograde planets. These objections are not too serious since the probability of all prograde orbits is not vanishingly small and evolutionary factors could create a near planar system from one much less so.

3.5   NEW NEBULA IDEAS

There have been a number of attempts to resuscitate the idea of the coeval formation of the sun and planets from a nebula. Such ideas are usually formulated and then abandoned within a short time but, to illustrate the complexity of the ideas which are deployed, a model given by Cameron in 1978 will be described.

The scenario begins with a high-density cloud of mass about $10\,000M_\odot$ which is somehow induced to collapse. Cameron suggests that this may have been by the action of a nearby supernova, an event which might also explain some of the isotopic anomalies found in meteorites (section 13.2). When the cloud collapses, it fragments and a primitive solar nebula, with mass $2M_\odot$, radius 100 AU and temperature 5–10 K, separates out.

The collapse of the nebula will eventually lead to a central condensation surrounded by a disk, and Cameron called on a theory given by Lynden-Bell and Pringle (1964) to describe the subsequent evolution of the central mass and the disk. This theory notes that if there is friction between different layers of the disk then energy must be lost but, at the same time, angular momentum must be conserved. The way for this to happen is for material near the spin axis to move inwards while that further out moves outwards. The net effect is an outwards transfer of angular momentum. An important requirement for this mechanism to operate is that neighbouring layers of the disk should be heavily coupled. Cameron associates this coupling with three agencies which stir up the disk. The first of these is due to the fact that in a rotating fluid circulation currents — small eddies — are set up spontaneously. The second agency is material which falls in towards the plane of the disk under gravity and which does not match in angular speed the material with which it mixes. Finally, in a very massive disk, condensations will form and the motion of condensed bodies through the disk will contribute to the stirring effect.

Material falling inwards on the disk would be rapidly decelerated and would produce a coronal (high-temperature) layer at its surface. The tendency for this hot material to be lost would, at first, be counteracted by the ram pressure of impinging material, but when this latter material became exhausted a mass-loss regime would become established.

While the disk was still quite massive a series of rings would form in it, each ring having about one per cent of the mass of the sun. In these rings condensations would form, much as they did in Jeans' tidal filament, and an accumulation of these would lead to protoplanets with masses several times that of Jupiter. Those close to the sun would lose all their volatile material while those further out, although losing much mass, would retain considerable gaseous and volatile components.

Finally Cameron ascribes the origin of satellite systems to the formation of rings in a disk which would have formed around the developing protoplanets.

This model does not solve the problems which have beset previous nebula theories. Protoplanetary condensations are formed but only because so much mass was originally present and the difficulty of disposing of the excess mass was mentioned in section 3.3. The Lynden-Bell and Pringle mechanism, while it does transport some angular momentum outwards, does little to solve the angular momentum problem. It leads to a massive inwards transfer of material which is momentarily in orbit about the growing central body before being absorbed and we saw, when the Poynting–Robertson effect was discussed, that this would give far too much angular momentum to the central body.

Another mechanism which has been suggested for the removal of angular momentum from the sun is the action of a magnetic field to transfer angular momentum from the core to the disk. An idea by Hoyle in 1960 envisages that a gap developed between core and disk and that the two parts of the system were linked by a magnetic field which traversed ionized, and hence conducting, material at the edge of the disk. If this ionized material had a high enough conductivity, then the magnetic field could be frozen in, that is to say that the magnetic field lines are constrained to move with the material. The pattern of field lines linking the core and disk is shown in Fig. 3.12(a). If the core now shrinks under the influence of self-gravitational forces then it attempts to rotate more quickly. The lines of flux are then stretched, since they are anchored at each end, they act like rubber bands under tension and this has the effect of pulling back on the core, so slowing down its rotation, while at the same time making the disk rotate more rapidly (Fig. 3.12(b)). One aspect of this idea is that if it is to work then the field associated with the early sun would have to have been much stronger than the present solar field. However, a more important criticism is that the magnetic field pattern shown in Fig. 3.12(b) is unstable; the magnetic field lines would break and reform to give shorter lines of flux.

Several ideas have been advanced for the loss of angular momentum of the sun due to the action of a magnetic field. One of these also appeals to T-Tauri emission by the early sun. It is assumed that charged particles emitted by the sun would be attached to radial magnetic field lines by spiralling around them and so, out to a certain distance, would co-rotate with the sun. This material would thus be increasing its distance from the sun while its angular velocity remained constant; consequently its angular momentum would increase at the expense of that of the central body which would therefore rotate more slowly. The idea is quite attractive but, once again, when reasonable numbers are inserted in the equations it fails by a considerable margin to extract enough angular momentum from the central body.

During the 1970s the solar nebula concept became established as a fundamental assumption of astronomy, notwithstanding that its two-hundred-year-old problems had not been resolved. Among the most faithful adherents were meteoriticists who have sought to explain features of meteorites in terms of condensation sequences from a hot solar nebula. This support was given despite the fact that Cameron, in a

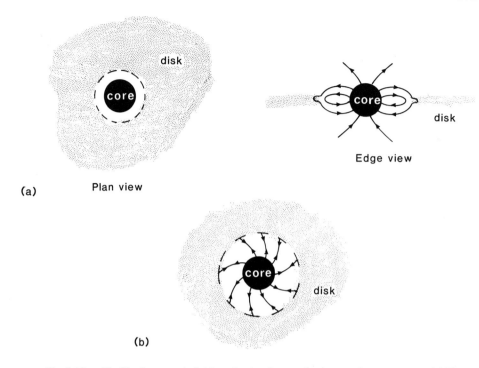

Fig. 3.12 — The Hoyle magnetic field mechanism for transferring angular momentum. (a) The magnetic field pattern. (b) Distortion of the field as the core collapses. The arrows show the directions of the forces on the core and the disk.

revision of his original ideas, has stated quite explicitly that at no time during the evolution of a solar nebula were temperatures reached which could have vaporized silicate materials.

Finally we may note that one difficulty common to all solar nebula theories concerns the rotation axis of the sun, which is at 7° to that of the system as a whole. It is not feasible that the rotation axis of the central body could be so inclined to that of the disk or, alternatively, that planets produced within the disk could systematically depart so much from its plane.

4

What a theory should explain

4.1 ESSENTIAL PROBLEMS FOR THEORIES

In considering the essential problems to be solved by a would-be theory of the origin of planetary systems in general, the assumption has to be made that some of the first-order, gross features of the solar system are characteristic of all planetary systems. This assumption really begs the question of whether other planetary systems exist at all and, if so, whether they resemble one another to the extent that the solar system may be taken as a good representative of all of them. Later we shall look at the evidence for and against the existence of other planetary systems, but for the present we shall just accept the current belief that the solar system is not unique and that other planetary systems do exist.

Looked at with the lowest resolution, the solar system is seen to consist of the sun, with the characteristic slow rotation of a late-type star, accompanied by a family of planets in direct, more-or-less coplanar orbits. From this, and chemical knowledge of our own planet, the following are suggested as the *essential* features to be explained by successful cosmogonic theory:

(1) the angular momentum distribution between the planets and sun;
(2) a mechanism for planet formation from whatever material is available;
(3) a cold origin of planetary material;
(4) direct and almost coplanar orbits for the planets.

Of all these features, that of the distribution of angular momentum seems to be the most demanding. In their different ways neither the nebula nor the tidal theories could resolve this problem. For the nebula theory there is no difficulty in producing material in rings with the proper intrinsic angular momentum; here the problem is that the sun is formed spinning much to quickly, if it is formed at all. The tidal theory, with the sun in a highly condensed state, sidesteps the problem of the slow rotation of the sun by presupposing its existence as it is now, but then there seems to be no way of imparting sufficient angular momentum to its material to form planets sufficiently far out. On the other hand, if the sun is presumed to have been in a very diffuse state, then potential planetary material can be produced with the right amount of angular momentum, but computational models show that in the tidal process, the sun would acquire at least as much angular momentum as all the planets combined (see section 8.2 and Table 8.3).

Of the more recent theories, there are two giving a partial or complete explanation of the angular momentum problem. The Schmidt–Lyttleton mechanism, by which the sun captures a nebula from an interstellar cloud, explains well how orbiting material with the mass of the planets could have had the requisite angular momentum. McCrea's floccule theory purports to do even better and explains both the slow rotation of the sun *and* the angular momentum associated with the planets. However, if the argument based on Fig. 3.11 is correct then the planetary angular momentum may still not be explained by the floccule theory.

The problem of forming the planets themselves must surely be very central to a proper theory of origin of planetary systems, and here too the position is far from satisfactory. In Laplace's nebula theory the planetary material is left in orbit in a diffuse state and the mechanism of its accretion to form larger bodies is described in vague non-quantitative terms. By contrast the tidal theory does initially organize material into protoplanetary blobs within the tidal filament and, if their temperature is low enough, they could condense. The floccule theory tackles the planetary formation problem by the supposition that the planetary material is organized into cold dense bodies, but even these have a lifetime too short to allow them to collect together to form a mass greater than the Jeans critical mass.

An outstanding feature of the solar system is its approximate planarity, but this is not a very demanding requirement and is satisfied by both of the older theories. The plane of rotation of Laplace's disk and the plane of the star–sun orbit in the tidal theory readily explain this characteristic. However, it should not be thought that a plausible theory of planetary formation requires that planets should be created in near-coplanar orbits; it would be enough to show that, even if the initial system was not coplanar, it would somehow evolve to that state. The condition that the original orbits should be prograde seems to be more stringent; if bodies are orginally produced in retrograde orbits then some evolutionary process must be invoked to remove them since it is unlikely that any mechanism could change their direction of motion around the sun.

It is a sobering thought that, for all the ingenuity and effort which have been expended on the theories described thus far, not one of them adequately explains all the features (1) to (4) given above. In other words, not one of them is capable of giving a reasonable description of the origin of planetary systems sufficient to explain even the crudest first-order structure of the solar system.

4.2   IMPORTANT PROBLEMS FOR THEORIES

One of the criticisms levelled at Jeans' original tidal theory was that it depended on an event that was so unlikely that the solar system might be the only planetary system in the galaxy or even the universe. In the context of its time, as an argument against the tidal theory, it had no logical basis. Whether the solar system is unique or not, it provides the environment for intelligent life to evolve and to debate the question of its uniqueness and there was no evidence to suggest that this matter should be under consideration anywhere else in the universe!

Notwithstanding the illogicality of the original criticism made against Jeans' theory, there is nowadays more reason to ask that a theory should predict that planetary systems are not too rare. It is claimed that observational evidence exists that some nearby stars, e.g. Barnard's star and Epsilon Eridani, have planetary companions. The validity of the evidence has been hotly disputed, depending as it does on observations of oscillations in the proper motions of stars at, or many say beyond, the limits of what can be achieved with optical telescopes. However, it has also been claimed that planet-sized objects, perhaps a few times as massive as Jupiter, have been detected in association with some stars through their infra-red emission picked up by IRAS (Infra-red Astronomical Satellite). Such interpretations might be open to the objection that what were being detected were binary systems for which one partner was not quite massive enough to initiate nuclear reactions and that a planetary *system* requires several small companions in an ordered arrangement. Given this situation a sensible stand to take is that preference should be given to a theory predicting a multiplicity of planetary systems; a theory which makes the solar system unique by appealing to some extremely unlikely cosmic event could not automatically be excluded but should, at least, be suspect.

A possible, but not infallible, way of deciding which are the features of planetary systems in general is too look for those features of the solar system which appear in a characteristic form in widely separated parts of the system. One such feature is that the planets Jupiter, Saturn and Uranus possess families of regular satellites. Associated with each of these satellite families there is a ring system, although the rings differ in character from one planet to another.

The above considerations lead to two further features which should be explained by a convincing theory of the origin of planetary systems, although they are less critical than those given in section 4.1:

(5) regular satellite systems should be formed with several, if not all, the planets;
(6) a theory is preferred if it predicts the existence of many planetary systems.

With the exception of the Jeans tidal theory, all the others which have been described would give a high frequency of planetary systems. The nebula and floccule theories envisage a planetary system as part of the same process which produces a star while Lyttleton claims that the passage of a star through an interstellar cloud, giving rise to the capture of a nebula, is likely to happen to any star during its lifetime.

For the formation of satellite families the floccule theory calls on the idea of droplet formation between the two parts of a rotationally disrupted protoplanet. With the nebula theory, as given by Cameron for example, an evolving planet becomes a small-scale version of the original nebula with satellites forming in the circumplanetary disk. The same concept, that satellites were produced by a scaled-down version of the process which gave planets, was also favoured by Jeans. Indeed, in 1929 Jeans went on record with the statement 'Each of these small systems is so exact a replica in miniature of the solar system that no suggested origin for the main system can be accepted unless it can account equally for the smaller planetary systems; any hypothesis which assigned different origins to the main system and the sub-systems would be condemned by its own artificiality'. Jeans then went on to describe how a newly formed planet, by interaction with the sun or passing star or both bodies together, would produce a tidal filament within which protosatellite condensations would form.

4.3   PROBLEMS FOR THE SOLAR SYSTEM

The six problems described in the previous two sections are all those which could reasonably apply to a general theory of the origin of planetary systems. However, within the solar system itself we find a number of detailed features which are probably due to the particular way in which the solar system evolved. A general theory for planetary formation must allow such evolution to take place and thus particular features of the solar system may impose some constraints, albeit weak ones, on a general theory.

As an example we may reconsider condition (4) which specifies that planetary orbits should be nearly coplanar. In fact the system has significant departures from coplanarity. The angular momentum vectors associated with the spin of the sun and the orbits of the planets are far from parallel (at 7° to each other) and, if the spin axes of the planets are included, then there is an even greater spread of directions. To put the 7° in context it should be noted that the probability of two vectors having this angle or less between them, just by chance, is about 0.004. Any theory which led to strictly parallel vectors would need to introduce a mechanism to destroy the parallelism — yet the theory must give near parallelism.

Another notable feature of the solar system is the spacing of the planetary orbits following a regular progression, indicated by Bode's law, out of Uranus. The law breaks down for Neptune, so perhaps not too much should be made of the precise form of the law, but a theory should give something like the observed spacing or suggest a mechanism for evolving to a similar pattern.

The decision which led to condition (5), which was taken to apply to planetary systems in general, specified only *regular* satellites as observed for Jupiter, Saturn and Uranus. There are also a large number of *irregular* satellites in the system. Two of these are large bodies, the moon and Triton but most of them are relatively small bodies with dimensions of order tens to hundreds of kilometres. Several of these irregular satellites, including the aforementioned Triton, have retrograde orbits and most of the orbits are well out of the plane of the planetary equator and have fairly large eccentricities.

For the rocky bodies in the central part of the solar system, Mercury, Venus, the earth, the moon and Mars, densities are only poorly correlated with mass (see Tables 2.2 and 2.3). Thus Mercury, which has half the mass of Mars, is as dense as the earth; intrinsically even denser if compressional effects are taken into account. While McCrea has paid some attention to these bodies, excluding the moon, in describing their origin in terms of the rotational disruption of protoplanetary cores, it cannot be said that the mass–density relationship has been explained at all.

For all the solid bodies of the solar system, as far out as we have observed, there is evidence of extensive bombardment damage. During the stage at, or just after, the formation of the solar system there appears to have been a period with a high density of orbiting debris. It seems reasonable to think of the asteroid belt as that residue of the debris which happens not to have interacted with other large bodies and so has persisted to the present time.

The only other small body, not mentioned above, is Pluto, with a mass about one-fifth that of the moon. It is characterized by having an orbit which is highly eccentric ($e = 0.249$) and inclined at 17° to the general plane of the system. The inner part of this orbit is just inside the orbit of Neptune and there may be an association between these two planets. For all its small size, Pluto is also distinguished by possessing an even smaller satellite companion.

One final feature of the solar system, which is probably not a feature of all planetary systems, is the existence of meteorites (and asteroids) with the properties we observe. A satisfactory theory of the evolution of the solar system ought to be able to account for the major features of meteorites as described in detail in section 13.2. These include evidence of temperatures in the early solar system high enough to have vaporized silicate materials. In addition there are measured isotopic compositions for some common elements, such as oxygen, magnesium and neon, which are anomalous compared with those measured for terrestrial samples. Such measurements and observations must give important clues about some aspect of the development of the system.

To summarize these points the following problems and properties of the system are suggested for the attention of any would-be theory of solar-system origin and evolution:

(7) the departure from planarity of the solar system;
(8) the progression of planetary orbital radii;
(9) the existence and properties of irregular satellites;
(10) the mass–density relationship, or lack of it, for the larger bodies in the terrestrial region;
(11) the origin of the projectiles which produced damage features on bodies throughout the solar system;
(12) the origin of Pluto and the characteristics of its orbit;
(13) the structure and composition of meteorites.

4.4   A STATEMENT OF OBJECTIVES

In selecting and ranking problems in order of importance there is inevitably a subjective element. An independent selection might be somewhat different but would necessarily heavily overlap the selection made here. It is certainly reasonable to say that any theory which could not provide explanations to the problems posed in section 4.1 would have no claim for acceptance. Equally, at the other extreme, any theory giving plausible explanations of all 13 points in a logical way would deserve careful attention.

In the following chapters there will be described, systematically and in logical sequence, a theory which explains all the thirteen points and problems and also other, less critical, ones. Most explanations are very detailed, others less so and rather more qualitative, but in no case will the explanation be other than in terms of simple mechanisms and phenomena which may be readily understood. The plausibility of the theory must be judged either in the light of the criteria suggested here or, alternatively, against any other acceptable and sensible criteria.

Any theoretical work in a field such as this must involve speculation, but speculation can be at various levels. For example, to speculate that physical constants may vary with time is a hypothesis for which there is no corroborative evidence — although eminent and respected scientists have so suggested from time to time. Yet another type of speculation may concern, for example, the evolution of the magnetic field of the sun. To explain the magnetism of lunar rocks it has been suggested that the magnetic dipole moment of the sun was higher when it first formed. There are theoretical reasons for believing that this might have been so; magnetism of stars is thought to be related to mass motions within them and a young star, still evolving quickly, would be a fairly turbulent object. Finally, as a consequence of a scenario under consideration, it might be concluded that, within the framework of well-understood physical laws, some event might have taken place — a tidal interaction or collision between two bodies, for example. It is this last kind of speculation, the most acceptable in terms of credibility, which is at the heart of the theory to be described.

The description of this theory may sometimes appear to be expressed in the language of certainty and speculative statements made with undue confidence. In excuse for this it must be said that the constant use of cautious terminology and the incessant repetition of qualifying and conditional phrases is even more tedious for a reader than for an author. Actually, in a field of this kind, there is no such thing as a *correct* theory. Theories are of two kinds — those that are plausible and those that are wrong! Any worthwhile theory must be detailed enough to be vulnerable so that some conclusion from it is capable of being refuted by observation or theory. That means that any theory with the status of being plausible may at any time be demoted to the second category of theories — those that are wrong. This may be the eventual, perhaps even inevitable, fate of the theory described here.

5

Star formation

5.1 STAR FORMATION AND PLANETARY SYSTEMS

There are those who would argue that it is unreasonable to consider the origin of planets without at the same time considering star formation. This view, upheld by Hoyle and McCrea, for example, is not without a rational basis. It is clear that, whatever the intervening processes, the initial raw material for stars is the interstellar medium which pervades the whole galaxy. This material is at a very low density $(10^{-21}$ kg m$^{-3})$ with kinetic temperature of order 10 000 K and it rotates with an angular speed characteristic of that of the galaxy $(10^{-15}$ radians s$^{-1})$. If a mass of such material were to collapse to form a star then it would rotate more rapidly so as to conserve angular momentum, much as an ice-skater spins more rapidly as he brings in his arms towards his body. On the basis of this argument it is readily shown that the sun would be rotating at about two revolutions per second, corresponding to an equatorial speed of twenty times the speed of light!

The above argument rests on the assumption that the material collapsing to form the star is completely isolated but, if it was coupled to the outside medium in some way, than its rotation may be reduced with angular momentum being transferred to external material. Hoyle has suggested that some form of magnetic coupling between the collapsing cloud and the interstellar medium could provide the necessary braking effect. There is in fact a very weak galactic magnetic field of some 10^{-10} T (one millionth of the Earth's surface field) but this is far too feeble to influence the motion of matter which has more than about twice the interstellar density. However, Hoyle (see section 3.5) and Alfvén and Arrhenius (1975), amongst others, have suggested that the magnetic field of the evolving star itself would be available to provide coupling with the outside medium. This scenario presents two difficulties: firstly, and depending on the type of mechanism being proposed, it would require magnetic fields some 100 to 10 000 times stronger than the present value to explain the slow rotation of the sun and, secondly, the presence of a magnetic field in the early stages of collapse would prevent the collapse and hence the production of a suitable field. It is a classic chicken-and-egg situation.

One important argument which has been advanced to support a link between star and planetary formation is based on the observations of stellar rotations (Fig. 5.1). The so-called late-type stars, with masses less than about $1.35M_\odot$, tend to have low equatorial speeds; for the sun this speed is about 2 km s^{-1}. For stars with mass greater than $1.35M_\odot$ the equatorial speed increases quite rapidly with mass, reaching a peak of a little over 200 km s^{-1} at $10M_\odot$ and slightly declining thereafter. An interpretation which has been made by some from these observations is that the late-type stars possess planetary systems which are formed as an intrinsic part of their evolutionary pattern so that the central star, which is all that we can actually observe, is denuded of angular momentum. Such a hypothesis would make planetary systems very common and within the past few years attempts have been made to detect the presence of planets around some nearby stars (section 4.2). For example, Van de Kamp has explained fluctuations of the proper motion of Barnard's star in terms of the presence of one or more planets of Jupiter-like mass. These, and other observations of a similar kind, have been heavily criticized and the direct observational evidence for other planetary systems is, at best, very uncertain.

A more indirect argument for the existence of planets has been put forward by Abt (1977). He has found that for binary systems there is a systematic variation of frequency of occurrence with mass ratio. Stars of mass less than about $0.07M_\odot$ will not be self-luminous and Abt assumes that the cut-off in observed mass ratio is due only to observational difficulty with low-mass bodies. By extrapolation from the frequencies of observed binary systems down to the mass ratios appropriate to planetary systems, Abt concludes that virtually all stars belong to multiple systems and that many of them will have planetary families. Some doubt must exist about the validity of the assumptions on which Abt bases his argument. While there may be many systems with brown-dwarf members, unobservable in visible light, there is a considerable difference between simple star systems and a highly complex planetary system such as the solar system.

5.2   STELLAR CLUSTERS AND STAR FORMATION

Observation shows that vast numbers of stars are organized into clusters which are of two distinct types. First there are the *globular* clusters, containing 10^5–10^6 stars and distributed in a halo around the galaxy, most of them well away from the galactic mean plane. They contain old Population II stars with low metal abundances and which, because of their age, have little to tell us about star formation. If one is to learn something about the birth of stars then the best stars to study are those which are young and may still retain some characteristics of the birth process. The second type of cluster, the *galactic* or *open* cluster, is much smaller with 100–1000 member stars; such clusters tend to be close to the plane of the galaxy and they contain comparatively young Population I stars. The reason why the stars are young is that the lifetimes of the clusters are short: in the range 10^8–10^9 years. Interactions, and hence energy exchange between the stars, are frequent and occasionally a star receives enough energy to escape from the cluster, thereafter continuing its existence as an isolated field star. Eventually all clusters are dispersed in this way, the final residue being a small stable system, possibly a binary system. Very old stars within galactic clusters are rare, simply because the clusters do not usually survive long enough to contain them.

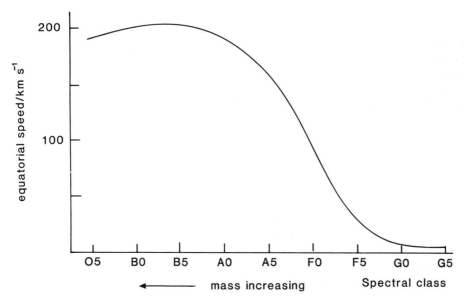

Fig. 5.1 — Variation of average equatorial speed with spectral class of star.

While it cannot be stated with certainty that all stars originate in clusters it is certain that many of them do so. There is something conducive to star formation in associations and we are going to examine a plausible model for the formation of a galactic star cluster.

5.3   EQUILIBRIUM IN THE INTERSTELLAR MEDIUM

The starting point for a galactic cluster is one of the dense clouds whose presence is most evident by observation, since they blot out the light from stars behind them in the line of sight. In these clouds the hydrogen, like most other elements, is in the form of molecules; while H_2 is difficult to detect directly, it can be detected indirectly by the observation of spectral lines from carbon monoxide excited by collisions with hydrogen molecules.

It may be deduced that such clouds may have masses from under one thousand to a million or more times a solar mass, densities of order 10^{-19}–10^{-20} kg m^{-3} and temperatures in the range 5–50 K. To understand how a dense, cool gas cloud forms it is necessary to consider the mechanisms by which heating and cooling processes take place in interstellar material. The main external sources of heating are background starlight and cosmic rays, the latter probably being the more important. Where the medium is transparent, each individual atom, molecule and dust grain in the cloud is bathed in the impinging radiation and the energy absorbed per unit mass of the cloud, which is proportional to the heating rate, is independent of the density or temperature of the cloud material. On the other hand, cooling processes are highly density- and temperature-dependent. The processes which occur are:

(i) Grain cooling

This kind of cooling was first described by Hayashi (1966). The mechanism assumes that the solid grains in the cloud are at a constant temperature, θ_g, and that a gas molecule striking the grain with a speed characteristic of the general temperature of the cloud material, θ, will leave with a speed appropriate to θ_g. The cooling rate will obviously depend on the number, density and dimensions of the grains, which determine the total target area and hence the rate of molecule–grain collisions. There is also a strong temperature dependence of the cooling; the higher the temperature, the faster the molecules move and the more often do they collide with grains. In addition the mean energy loss per collision depends on $(\theta-\theta_g)$ which also increases with θ. The cooling rate is also dependent on density in quite a simple way. If, for example, the density is doubled then the collision rate will increase by a factor of four — in each unit volume there are twice as many molecules making twice as many collisions per unit time. The cooling rate per unit volume thus increases by a factor of four, but within this volume there are twice as many molecules from which the energy is taken. Therefore the cooling rate per unit mass doubles when the density doubles and it is readily seen that the cooling rate per unit mass is directly proportional to the density.

(ii) Excitation by electron or atomic collisions

Within a cloud there will exist a number of atomic ions — atoms which have lost one or more electrons by interaction with external radiation of some kind. Seaton (1955) has considered the way in which the presence of the ions C^+, Si^+, Fe^+ and Mg^+ contribute to cooling. Free electrons in the cloud will collide with these ions and give up some of their energy to a bound electron which will be promoted to a higher energy state. Subsequently the electron will fall back to its original state, giving off a photon which, because the cloud is transparent, will simply leave the system. The net result of this process is that part of the energy of the original colliding electron has been lost from the cloud. Since the electron energy of motion represented part of the total thermal energy of the cloud material, this amounts to a cooling process. Seaton has given an expression for the cooling rate per unit mass based on the expected population densities of the kind of ions mentioned above. The dependence on temperature is fairly complicated but, as for dust cooling, the loss of energy per unit mass is proportional to the density.

Another way in which energy can be lost is by the excitation of either atomic oxygen or molecular hydrogen by atomic collision. In the case of hydrogen the molecule can take up energy in the form of rotation; when the molecule falls back to a state of lower rotational energy then, again, a photon is emitted and energy is lost from the system.

The variation of the cooling rates due to these various mechanisms for a particular cloud constitution is illustrated in Fig. 5.2. The discontinuity in the grain cooling curve is due to volatile grains evaporating at a temperature of about 120 K; thereafter, grain cooling is mediated by silicate and metal grains alone. The great temperature-dependence of the cooling mechanisms is very readily explained. For example, to cause a hydrogen molecule to rotate with the lowest possible energy it is necessary for the colliding particle to have energy appropriate to a temperature just

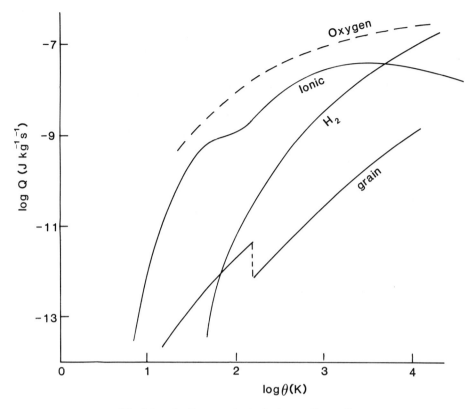

Fig. 5.2 — Cooling agencies in the interstellar medium.

under 100 K. For a particle with less energy (speed) than this the hydrogen molecule cannot be sent into a tumbling mode and the cooling process will not take place at all.

Since there is a density, ρ, and temperature, θ, dependence of cooling it is possible to find combinations of (ρ,θ) all corresponding to the same cooling rate. However, pressure, p, also depends on these two quantities (it is proportional to the product $\rho\theta$) so that it is also possible to find combinations of (p,ρ) corresponding to the same cooling rate — which is more convenient for our present purpose. The type of relationship between p and ρ for a particular rate of cooling is shown by the full line in Fig. 5.3 where, because of the large range of p and ρ, these quantitites are plotted on a logarithmic scale. A feature of interest is illustrated by the horizontal line ABC, representing a fixed pressure. The points A, B and C correspond to three values of density, or temperature, for which both the pressure and the cooling rates per unit mass are identical. Actually point B corresponds to an unstable state since the curve at this point gives pressure decreasing with increasing density, so only the points A and C can exist in practice. From this we see the way in which dense cool clouds can exist in or near equilibrium in the interstellar medium. Point A corresponds to a low density and a high temperature (approximately 10^{-21} kg m^{-3} and 10000 K) while point C represents a relatively high density and low temperature (approximately 10^{-18} kg m^{-3} and 10 K). A region with conditions corresponding to point C can be in

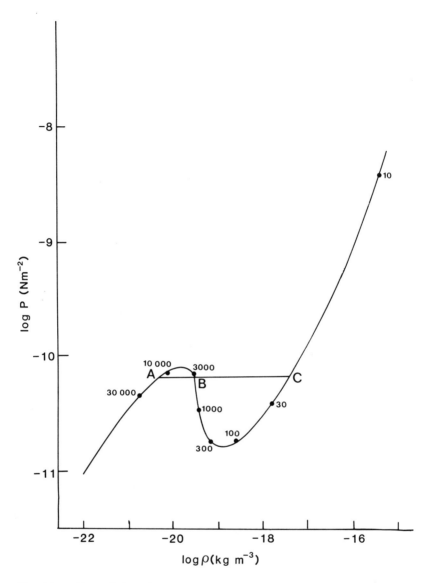

Fig. 5.3 — The variation of pressure and density to give a particular cooling rate in an interstellar cloud. Numbers on the curve indicate temperature.

pressure equilibrium with a surrounding region with conditions corresponding to A, while both regions can be in thermal equilibrium with the cosmic ray input of energy, which is of order 10^{-8} J kg^{-1} s^{-1}.

5.4 THE FORMATION OF DENSE COOL INTERSTELLAR CLOUDS

Although it has been shown that cool dense clouds can exist in a state of equilibrium within the general interstellar medium, the question must be addressed of how they come to form in the first place. One way for this to happen is to inject extra solid or high-atomic-mass material into a region of space. The effect of this can be followed in Fig. 5.4. Since the heavy-element component of interstellar space is only about one

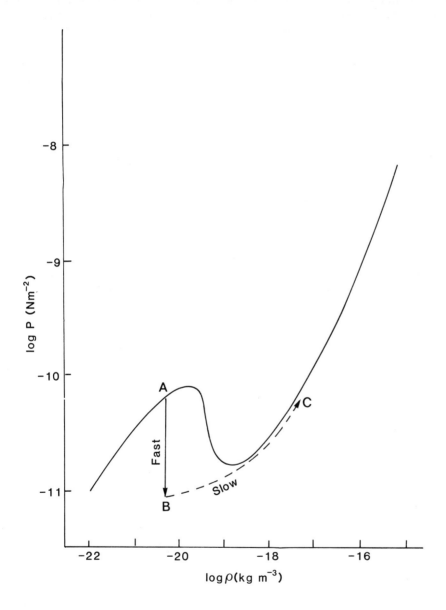

Fig. 5.4 — The effect of injecting coolant into an interstellar medium. The change of conditions is represented by the path ABC.

per cent by mass, even doubling the amount of non-hydrogen coolant material would make very little difference to the overall density. However, with considerably more coolant present, the temperature required to be in equilibrium with cosmic-ray heating would be lower. Cooling, which depends to a large extent on the speed of electrons and on the radiation of energy, would take place much faster than changes of density, which depend on the mass motion of the material. Thus the effect of adding coolant material would be to change the state of the interstellar material from point A to point B on a timescale short by astronomical standards. However, the resultant reduction of pressure in the region means that it is no longer in equilibrium with surrounding material; there will be an influx of material increasing the pressure and density and, incidentally, promoting further cooling. The path in Fig. 5.4 would be B to C, the final point being close to, but not quite on, the original curve because of the added coolant, the effect of which would eventually be somewhat diluted by the influx of normal interstellar material. This scenario is quite plausible but one can only speculate about how the extra coolant material would arise. A strong possibility is through the injection of material due to a nearby supernova explosion. Indeed the idea that star or planetary formation is connected with a supernova event has been suggested by Cameron, as was mentioned in section 3.5, although for a different reason.

The above discussion has been concerned only with thermal and pressure equilibrium and any possible gravitational effects have been ignored. The concept of the Jeans' critical mass (equation (1.2)) has previously been mentioned and it is clear that if the mass of the cloud exceeds M_J, given its density and temperature, then the cloud will spontaneously collapse under gravitational forces. For various reasonable starting sets of conditions, with the cosmic-ray heating rate given at the end of section 5.3, the density and temperature corresponding to point C in Fig. 5.4 give a critical mass of hundreds to thousands of times a solar mass; a very suitable starting minimum mass for the creation of a galactic cluster.

5.5 OBSERVATIONAL INFORMATION

All we can know about stars comes from the light they emit and this information is of two kinds. Firstly there is the absolute magnitude of the star, which is a measure of its luminosity or its energy output per unit time. Secondly there is its spectral class, which is observed in terms of colour and spectral lines, but is also a measure of the star's temperature. A plot of these two characteristics constitutes what is known as a Hertzsprung–Russell (H–R) diagram and from the position of a star on the diagram most of its physical characteristics may be inferred.

When stars are plotted on a H–R diagram it is found that many of them lie within a narrow band, shown in Fig. 5.5, called the main sequence. Such stars are producing their energy by hydrogen to helium conversion and a star of one solar mass would spend about 10^{10} years in that state in practically the same position on the H–R diagram.

For main-sequence stars one can relate the mass, radius and luminosity to the spectral class, and some typical values are shown in Table 5.1. It will be seen from the luminosity that for the early-type O and B stars the fractional rate of consumption of hydrogen is extremely high and so, paradoxically, the more massive is a star the

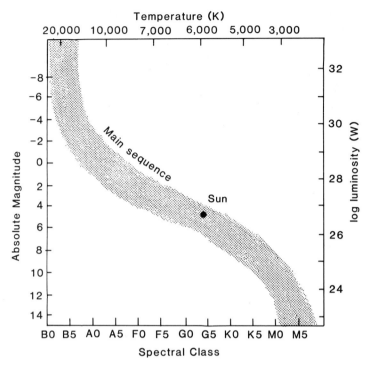

Fig. 5.5 — A Hertsprung–Russell diagram showing the main sequence band.

Table 5.1 — Masses, luminosities and radii of main-sequence stars in terms of solar values

Spectral Class	$M/M_\odot$	$L/L_\odot$	$R/R_\odot$
O5	39.8	3.2×10^5	17.8
B0	17.0	1.3×10^4	7.6
B5	7.1	631	4.0
A0	3.55	79	2.63
A5	2.19	20	1.78
F0	1.78	6.3	1.35
F5	1.41	2.5	1.20
G0	1.07	1.26	1.05
G5	0.93	0.79	0.93
K0	0.81	0.40	0.85
K5	0.69	0.16	0.74
M0	0.48	0.063	0.63
M5	0.22	0.008	0.32

shorter is its sojourn on the main sequence. Main-sequence lifetimes are given in Table 5.2; these are so short for O and B stars that such stars are comparatively rarely seen on the main sequence.

Table 5.2 — The lifetimes of stars on the main sequence

Mass/$M_\odot$	Lifetime (10^6 years)
15	10
9	22
5	68
3	230
2.25	500
1.5	1700
1.25	3000
1.0	8200

If we look at a cluster of stars which is less than a few million years old then all the stars, even the most massive, will be on the main sequence and the stars will all lie within the main-sequence band of the H–R diagram, as indicated in Fig. 5.6(a). However, after further time some of the very massive stars will exhaust most of their hydrogen fuel in their central regions and they then begin an evolutionary process which takes them off the main sequence and on a journey in the H–R diagram. This journey begins by motion to the right as the star expands and cools. Because the luminosity depends on the product $R^2\theta^4$, where R is the radius of the star and θ its surface temperature, it so happens that the absolute magnitude changes little. The star will become a red supergiant and, after 10^7 years, the H–R diagram of the cluster appears as in Fig. 5.6(b). The gap in the diagram corresponds to states of very short duration in which few, if any, stars will be found. As time goes on, so the point on the main sequence where stars move off to the right gradually moves downwards. The evolution away from the main sequence is slower for lower-mass stars and they become cooler but much more luminous as they move to the red giant region of the H–R diagram. The approximate appearances of the H–R diagram after 10^8 and 10^9 years is shown in Fig. 5.6(c) and (d). Later, red supergiant stars evolve leftward again on a journey which will eventually lead them to become white dwarfs at the bottom left-hand region of the H–R diagram. The H–R diagram for the old cluster M5 is shown in Fig. 5.7 and the various evolutionary features are well-represented in this diagram.

The above description of the evolution of stars *from* the main sequence is remote indeed from our objective of describing how stars are formed in the first place. We need now to consider the pre-main-sequence stages of stellar evolution, and a great deal of theoretical work has been done on this topic. While there are some differences of view in the details of pre-main-sequence evolution, there is a general agreement about the broad pattern. The form of evolution of a star of one solar mass,

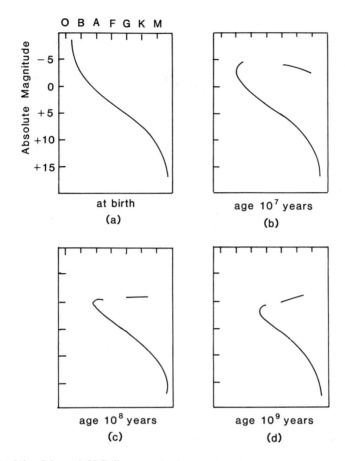

Fig. 5.6 — Schematic H–R diagrams of a cluster as it evolves. The turn-off point moves down as the cluster ages.

as given by Hayashi (1961), is illustrated in an H–R diagram in Fig. 5.8. The star begins at point A with a temperature of about 50 K and a very low density of order 10^{-15} kg m^{-3}. In view of its very low temperature and despite its very large surface area, it has less than one-half of the luminosity of the sun. As the star evolves towards B its temperature changes very little; although heat is produced by the gravitational collapse, the stellar material is so diffuse that it is transparent to the radiation produced within itself, which is radiated away. During this approximately constant temperature journey from A to B, the surface area, and hence the luminosity of the star, falls sharply. However, at the point B the material becomes more opaque and the energy of collapse is progressively less able easily to escape. This leads to a heating-up of the stellar material and a slowing-down of its rate of collapse as pressure gradients build up within it. There is a short duration (about 100 days) overshoot in the region C whereafter the star is in a state of near-equilibrium. As it radiates energy, so it slowly collapses to restore equilibrium and there is a gradual

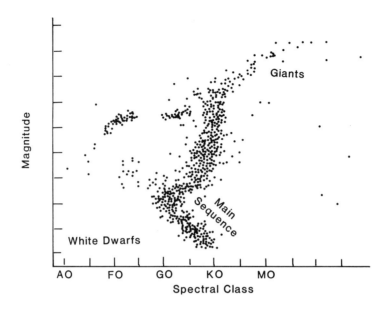

Fig. 5.7 — The H–R diagram for the globular cluster M5.

increase of energy as it moves towards the main sequence at D. This stage of slow collapse is called Kelvin–Helmholtz contraction.

Other calculations have been done which differ in detail from those of Hayashi, and the final stages in the evolutionary development of stars of various masses according to Iben are shown in Fig. 5.9. The total times, as a function of mass, for stars to reach the main sequence along the path of radiation-controlled slow collapse, as given by Iben, are given in Table 5.3.

It will have been noted that the paths of stars evolving towards the main sequence, and also away from it, are all on the same side of the main sequence band. Hence a star at a given point on that side could either be evolving towards or away from the main sequence. Actually there will never be any confusion for cluster stars. If the turn-off point on the H–R diagram indicates that the cluster is old, then all stars are either on the main sequence or moving away from it. On the other hand if the cluster shows O or B stars *on* the main sequence then the cluster must be very young and stars of lesser mass must either be on the main sequence or moving towards it. What is more, since the tracks in Fig. 5.9 do not intersect each other, then every point in the diagram occupied by these tracks corresponds to a unique mass and age for the young star. It would probably be more accurate to say *relative* age, for there is some arbitrariness in defining the time of birth of a star. A sensible birth time might be when the star became a distinct object in terms of having a mean density some number of times that of the background material. In what follows we shall be referring to the ages of stars based on a zero-age point defined by Iben and Talbot (1966). It is as well to realize that there may be some uncertainty here and that the ages so defined may not correspond to ages related to the concept of the 'distinct object'.

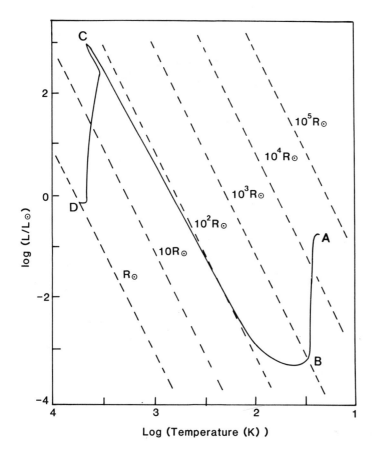

Fig. 5.8 — The path of a protostar towards the main sequence. The dashed lines give radii in solar units.

Based on the theoretical pre-main-sequence tracks, Iben and Talbot (1966) and Williams and Cremin (1969) have examined the masses and ages of stars in young galactic clusters — recognized by the presence of massive stars on the main sequence. In Fig. 5.10 there are illustrated the results of Williams and Cremin for the young galactic cluster NGC 2264. In Fig. 5.10(a) there is shown the mass–age relationship; it appears that the first stars produced have masses somewhat greater than one solar mass, and that at a later time stars both more massive and less massive are produced along two streams of development with a point of bifurcation some five million years ago. Derived from the same information, there is illustrated in Fig. 5.10(b) the evolution with time of the rate of formation of stars; the rate is slow at first but then builds up rapidly. Finally, in Fig. 5.10(c), there is shown the mass function, $f(M)$: the number of stars per unit mass range as a function of mass. For the higher-mass stars the relationship appears to be

$$f(M) \propto M^{-2.5} \tag{5.1}$$

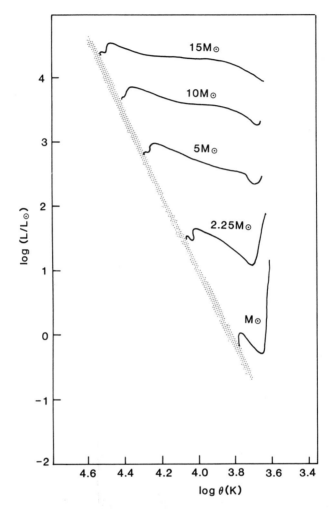

Fig. 5.9 — The slow collapse stage towards the main sequence governed by heat loss due to radiation.

which agrees quite well with the usually quoted mass index of -2.3 for stars in general. Williams and Cremin gave observational results for three other young clusters and these are similar to those from NGC 2264.

At the beginning of this chapter it was explained that one of the fundamental problems in star formation is to explain how stars could form from galactic material and yet be rotating so slowly. Unless during their evolution the stars interact with their environment in some way, their angular momentum will remain constant and, at some stage, they will be rotating so fast that they will no longer be able to collapse at all. Since the result of star–environment coupling is to reduce the angular momentum of the star, the present value of angular momentum must be a lower bound of the value at birth.

Table 5.3 — Times for model stars to reach the main
sequence along the radiation controlled path

Mass/$M_\odot$.	Time (10^6 years)
15	0.062
9	0.15
5	0.58
3	2.5
2.25	5.9
1.5	18
1.25	29
1.0	50
0.5	150

The rotational property of a star which may be directly inferred from observation is its equatorial speed. This is due to the Doppler shift in the wavelength of light seen by an observer when the source is moving. Thus if the source moves away from the observer the light is redder than it would be from a stationary source; conversely, if the source is moving towards the observer then the wavelength is shifted towards blue. The light seen by the observer from a star will have an average wavelength shift due to the motion of the star as a whole but, due to the spin of the star, some parts will be moving relative to the observer faster than average and some slower. For a particular spectral line there will be a shift of average wavelength corresponding to the average motion and a spread of wavelengths due to rotation.

If equatorial speeds are measured for stars of various spectral classes then the systematic variation is found which is illustrated in Fig. 5.1; it should be stressed that there is a considerable variation of equatorial speeds for stars of similar spectral class and what is shown is *average* speeds. What the figure shows is that for stars later than about F5 equatorial speeds are very low; for example, the sun, of spectral class G2, has an equatorial speed of only 2 km s^{-1}. For earlier-type stars the equatorial speed rapidly rises, reaching a peak somewhat over 200 km s^{-1} in the B5 region. It is interesting to note that the stellar mass in the region of the point of bifurcation in Fig. 5.10(a) also corresponds to about spectral class F5 and this suggests that such stars may have a special significance in the process of star formation. One interpretation which has been placed on the low equatorial speeds for late-type stars is that all such stars possess planetary systems. This argument suggests that the total angular momentum of a star, including any companions, would not be very variable but that the variation is only in that part of the angular momentum we are able to observe.

It is now well-established that stars form within dark cool interstellar dusty clouds in which much of the hydrogen is in molecular form. There have been recent measurements of hydroxyl and water maser sources within dark clouds which are almost certainly associated with star formation. The observations consist of extremely bright and sharp emission lines, for example at 0.18 m wavelength for OH, which contrast with the rather broad and shallow absorption lines which

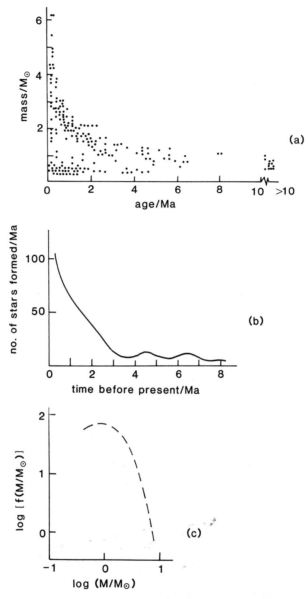

Fig. 5.10 — (a) Mass versus age for the cluster NGC 2264. (b) Rate of star formation for NGC 2264. (c) The mass-distribution function for NGC 2264.

astronomers are accustomed to seeing. The actual mechanism giving rise to maser emission is not known, although there are some tentative ideas about this. Hydroxyl maser sources come from regions of typical dimension 10^{14} to 10^{16} m with elementary sources, thought to be individually forming stars, of dimensions 10^{11} to 10^{12} m. In addition there are Doppler-shift-deduced velocities from various parts of the cloud of order 20 km s $^{-1}$ and this suggests the presence of considerable turbulence within the cloud.

With this observational background to guide us, we shall now develop a theory for the formation of a galactic cluster of stars and we shall see to what extent theory and observation can be matched.

5.6   THE COLLAPSE OF A CLOUD WITH TURBULENCE

Since the observational evidence suggests that dense turbulent clouds are the sites in which stars are formed, it seems intuitively obvious that it is by the collision of streams of turbulent material that matter may be concentrated to a suitable density for star formation. A model for the collapse of a turbulent cloud to give a glacatic star cluster (containing 100 to 1000 stars) has been given by Woolfson (1979). While the model is a somewhat simplified one — for example, it is assumed that the cloud remains homogeneous throughout the collapse — it does contain all the features one would expect in an actual collapsing cloud and it involves a considerable amount of mathematical analysis. What we shall do here is to describe the physical processes which underlie the analysis, give the equations from which the behaviour of the cloud may be deduced, and explain the meaning of the various terms in the equations. Finally the results of computational work using the equations will be given and compared with what we know or deduce from observation.

The starting point for the model is a spherical homogeneous cloud of radius R, mass M, temperature θ and radial speed of boundary material dR/dt. The mass motion of the cloud material can be divided into two components. The first of these is *linear-wave flow*, shown in Fig. 5.11, corresponding to a radial component of motion which preserves the form of the mass distribution within the body. The second component is *turbulent motion* which is a haphazard movement of material varying in speed and direction from point to point within the cloud and undergoing constant change. To an external observer of a turbulent medium there would be evident a number of characteristics of the system. One such characteristic would be the turbulent energy, as manifested by the mean-square average speed of the turbulent motion. This turbulent motion would be superimposed on any systematic motion of the medium such as the linear-wave flow previously described. Another characteristic would be a scale length, which would describe the dimension of a region in which the turbulent motion was significantly correlated; this is illustrated in Fig. 5.12. We shall refer to this scale length as the *characteristic length* of the turbulence.

Related to the mean-square speed of the turbulent motion and the characteristic length there will be a *characteristic time*. This will be such that two observations of the medium at an interval less than the characteristic time will show some correlation of the turbulence pattern. For observations separated by much more than the characteristic time there will be no obvious correlation. It seems reasonable to assume that the characteristic time, t_c, will be related to the characteristic length, l_c, and the mean tubulent speed , u, by

$$t_c \simeq l_c/u \tag{5.2}$$

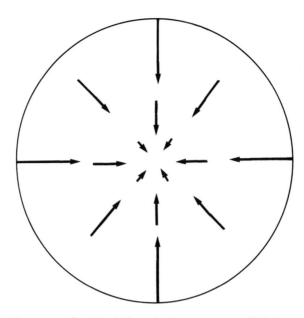

Fig. 5.11 — Linear wave flow in which each element moves radially inwards with speed
proportional to the distance from the centre and the symmetry of the body is preserved

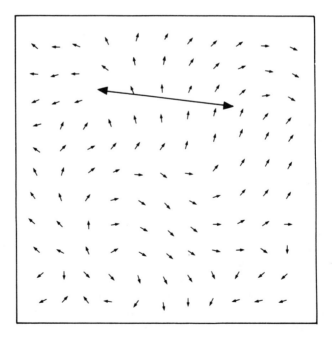

Fig. 5.12 — A representation of turbulent motion. The arrowed line gives an impression of the
characteristic length over which motions are correlated.

which is to say that by the time a turbulent region has travelled it own length its energy would have been redistributed into a new pattern of turbulence.

The first equation which can be derived gives the acceleration of the boundary of the cloud as

$$\frac{d^2 R}{dt^2} = \frac{5k\theta}{mR} + \frac{5\varepsilon}{3R} - \frac{GM}{R^2} L$$

(5.3)

where k is Boltzmann's constant, m the average mass of a gas molecule in the cloud and ε is twice the mean turbulent energy per unit mass of the cloud material (equal to the mean-square speed of turbulent motion). On the right-hand side the terms represent forces per unit mass on boundary material, the first being that due to thermal pressure, i.e. the temperature of the material; the second being due to turbulent pressure; and the final term the self-gravitational attraction of the cloud. The signs associated with these terms show that the first two are tending to cause the cloud to expand while the final term leads to contraction.

Next we consider the way that turbulence will develop in the cloud. The collapse of the cloud releases gravitational energy, part of which increases the temperature of the cloud material while the rest increases the turbulent energy. Turbulent streams of matter will constantly be slapping together with a consequent conversion of turbulent into thermal energy; clearly if the gravitational energy was not feeding the turbulence the latter would soon cease. The equation incorporating all these ideas, and giving the rate of increase of turbulent energy, is

$$\frac{d\varepsilon}{dt} = -\frac{2\varepsilon}{R}\frac{dR}{dt} - 2\,Q_t$$

(5.4)

where Q_t is the rate of dissipation of turbulent energy per unit mass of the cloud material.

Since for a collapsing cloud dR/dt is negative, the contribution of the first term on the right-hand side of equation (5.4) will be positive. The presence of the component ε in that term shows that in order to generate turbulence there has to be some turbulence present in the first place. However, by whatever means the dense cloud is formed — including by the injection of material from a supernova as previously suggested — some initial turbulence is bound to be present.

In order to make any progress with the investigation and analysis of a very complex phenomenon, and the collapse of a cloud with turbulence certainly qualifies for that description, it is necessary for the theorist to construct a model. The model must be tractable, otherwise there is no point in the exercise, but it must also include, in some form or other, all the important features of the real physical situation. The model used here to simulate a turbulent collapsing cloud is illustrated in Fig. 5.13. The whole volume of the cloud is envisaged as consisting of tightly packed spheres, each with a diameter corresponding to the turbulence characteristic length and each moving with the same speed, u, but in randomly chosen directions. The randomness of the motion causes the interaction of some neighbouring regions, which is the

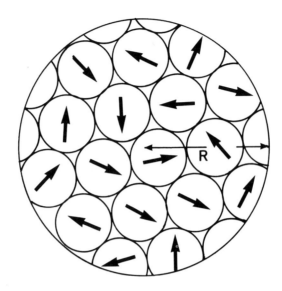

Fig. 5.13 — A simulation of a cloud with turbulence. Each spherical element has a radius R_J and moves at the same speed with respect to the centre of mass but in a random direction.

equivalent to the collision of turbulent streams in the real cloud. An assumption that has been made is that the characteristic length within the cloud is related to the Jeans radius, the radius corresponding to the critical mass in equation (1.2), and it has actually been taken as equal to a Jeans radius, R_J. This means that the sizes, and the masses, of the turbulent units depend on the density and temperature of the material, with any possible dependence on the turbulent speed being neglected.

An additional equation, which will not be given, gives the rate of change of the temperature of the cloud as it collapses. It turns out that the cooling processes described in section 5.3 are so efficient that the temperature changes little during the collapse. In addition, the overall behaviour of the cloud in terms of star formation varies little as a function of temperature. The nearly constant temperature regime breaks down in the final stages of collapse, when the cloud has become opaque to the radiation being produced within it, but by then most of the important events leading to star information have taken place. For these reasons, in the numerical simulations of the collapse of a turbulent cloud, the temperature has been kept constant.

5.7   THE PROCESS OF PROTOSTAR FORMATION

Since we are taking our turbulent elements to have a Jeans radius, then they will also have a Jeans critical mass and it might be thought that each turbulent element would spontaneously collapse to form a star. This is not so. A Jeans-mass unit will collapse on a timescale greater than the free-fall time appropriate to the initial density of the element; the free-fall time is the time taken under the forces of gravity alone for all the material to fall in to the centre. This time is

$$t_f = \left\{ \frac{3\pi}{32\rho G} \right\}^{1/2} \tag{5.5}$$

and the form of the collapse is such that there is very little change of dimension for a considerable proportion of the period t_f with a fast and accelerating collapse at the end. In the early stages of the collapse of the cloud t_f is very much greater than the characteristic timescale for turbulence, t_c, so that before a star can form the material is stirred back into the cloud.

From the above considerations it seems that no star can be produced until a region of appreciably higher density than average can form within the cloud. Such a region can be produced by the collision of turbulent elements.

In Fig. 5.14 there is depicted schematically the head-on collision of two superso-

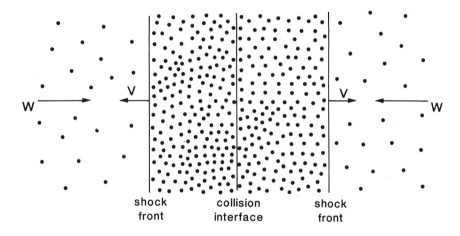

| shock | collision | shock |
| front | interface | front |

Fig. 5.14 — Compression and shock-front formation with colliding turbulent streams.

nic streams of gas. By supersonic we mean that the speed, w, with which the gas moves towards the collision interface is greater than the speed of sound in the gas, c. If the stream is supersonic then its Mach number, M $(=w/c)$, is greater than unity. The incoming gas meets a shockfront which is shown moving away from the collision interface with a speed v. The gas just behind the shockfront is both compressed and heated, and the degree of compression and the increase in temperature depend on the value of M. From the theory of supersonic collisions it is possible to make an estimate for Q_t in equation (5.4). It is assumed that the heated gas behind the shockfront cools back to its original temperature very quickly, due to the aforementioned efficiency of the cooling processes, and that it takes in heat again to maintain a constant temperature as it slowly re-expands. The difference between the heat radiated and the heat reabsorbed represents an energy loss of the material and the time during which this process takes place is of order t_c.

Returning to the compression of the material, it is clear that if M is too small then the compression of material will be insufficient to make $t_f < t_c$. On the other hand, for a head-on collision it is also possible to have M too large; there can be so much energy in the colliding streams that the gas is compressed into a disk, an unfavourable shape for gravitational collapse, and also given enough kinetic energy to overcome its own gravitational cohesion. However, even at such large values of M, star formation is possible for an oblique collision of two elements, which reduces the centre-to-centre relative speed. Taking all these factors, and others, into account it is possible to deduce a star-formation rate for any ρ, θ and ε and also to estimate the average mass of the stars being produced.

The results of a typical calculation are shown in Fig. 5.15 which shows the variation of R, ρ and u/c as time progresses. Features to be noted are:

(a) The value of dR/dt, the speed at which the boundary moves in, remains fairly constant. The accelerating collapse which would be caused by gravitational effects alone is opposed by the increasing turbulent pressure as the collapse proceeds.
(b) The Mach number of the turbulence, (u/c), steadily increases until, eventually, it is high enough to induce star formation.
(c) In Fig. 5.15(b) there are shown the number and range of masses of stars produced in each interval of 250 000 years. In these calculations the stars first produced at any significant rate are those with masses in the range $1.3–1.4M_\odot$.
(d) As time progresses so the rate of star formation increases, but there is a decrease in the masses of individual stars. Initially the calculated rate of formation agrees quite well with the deduced rate (Fig. 5.10(b)) but in the last one million years the rate rises up to four times that observed. Woolfson has explained this in terms of features which would be present in the very energetic final stages of collapse, including the exhaustion of material, but which have not been taken into account in the model.
(e) The mass index, illustrated in Fig. 5.15(c), of -2.6 agrees reasonably well with the observed value given in equation (5.1).

These calculations seem to reproduce quite well the lower branch of star formation as illustrated in Fig. 5.10(a). What is missing is the upper branch corresponding to the formation of stars with masses greater than about $1.35M_\odot$.

5.8   THE FORMATION OF MASSIVE STARS

The idea that is now examined is that the formation of more massive stars is due to accretion by the lower branch stars which have been formed previously. There are two types of accretion process which may occur. The first of these was described by Eddington (1926) and is illustrated in Fig. 5.16, where material moving relative to the star is attracted by its gravitational field, strikes the surface and adheres to it. The second type is the one where an accretion column is formed — the Bondi and Hoyle mechanism already discussed in section 3.2 and illustrated in Fig. 3.3. In this case the

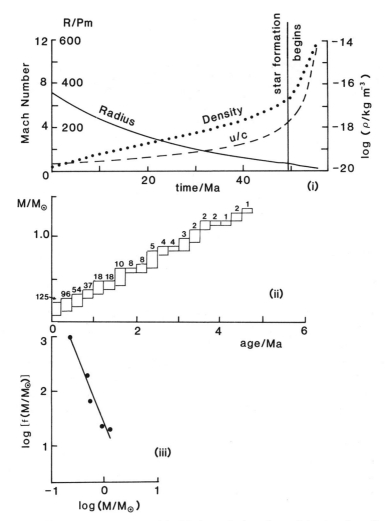

Fig. 5.15 — (i) Turbulent mean speed (as Mach number), radius and density of a collapsing cloud as a function of time. (ii) The number and masses of stars produced in 250 000-year intervals. (iii) The mass function for the stars produced.

accretion-column material is not captured around the central star, as previously assumed, but is actually accreted by it. In Fig. 5.16 there is shown the accretion radius, R_E, for the Eddington process; that is to say the greatest distance of the approach path of the material to the star centre for which accretion is possible. For a diffuse protostar, under the normal conditions to be expected, $R_E > R_B$, where R_B is the Bondi accretion radius given in equation (3.6), so that the former value would be the one to use. However, there are some other considerations which must be taken into account.

For a very diffuse protostar, say with mass $1.35 M_\odot$ and density 2×10^{-7} kg m^{-3}, the escape speed from the surface is about 340 m s^{-1}; not very different from the

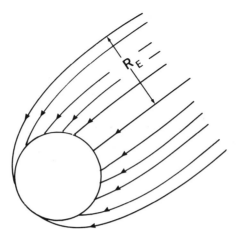

Fig. 5.16 — The Eddington accretion model where the stream of accreted material falls directly
on the protostar.

average speed of hydrogen atoms, or even molecules, at a temperature of 8 K. Thus, while the bulk of the star will be collapsing, there will be some evaporation from the surface, which will include any low-density accreted material. There will also be another effect which we can call abrasion. If the speed of the external medium relative to the star is much greater than the escape speed from the star, then accreted material, on striking the star, will share its energy with surface material and will not only immediately escape from the star but will also take other material with it.

The combined effect of evaporation and abrasion will depend on where and how the accreting material is striking the star and losses will be particularly severe where the accreted material strikes the surface at a high angle to the normal, say at point B in Fig. 5.17. The loss from a region such as A will be much less because there is a ram

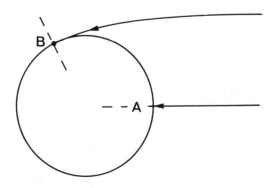

Fig. 5.17 — Streams of material falling at different points on the protostar do so at different
angles to the normal.

effect due to the impinging material. Taking these considerations into account it turns out that most accretion will take place when the protostar has partially collapsed and where R_B will be the appropriate accretion radius.

All accretion theories are based on the idea that the star is moving relative to a static quiescent ocean of gas — which is certainly not true in a turbulent cloud. Out to a certain distance R_T, which may be estimated in terms of the characteristic length in the turbulent medium, it will be approximately true, but the accretion radius to use in estimating the rate of gain of mass of the star is the lesser of R_T and R_B.

In the cloud-collapse model developed in section 5.6 it was assumed that the cloud remained homogeneous. This assumption is known to be invalid — although it will not invalidate the general conclusion concerning the pattern of collapse found in section 5.6. For example, in Fig. 5.18 there is shown the form of a collapsed cloud

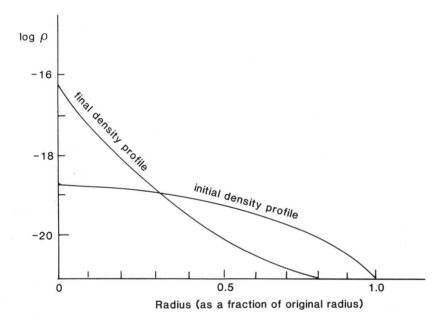

Fig. 5.18 — Schematic representation of the collapse of a gas cloud. After some time a density peak forms at the centre. The density in the central region is 100 to 1000 times the average for the whole cloud.

with a mass $5 \times 10^5 M_\odot$ found in an analysis by Disney, McNally and Wright (1969). It will be seen that the centre of the cloud may have density 100–1000 times greater than the average for the cloud as a whole and a star traversing the central region would have an accretion rate similarly enhanced. Based on various assumptions about the speed of the star relative to the medium, the degree of turbulence and the local density enhancement over the average density, mass-accretion lines have been calculated starting from various points on the lower branch of star formation and the results are illustrated in Fig. 5.19. It will be seen that the network of mass-accretion lines occupies a region similar to that occupied by stars in Fig. 5.10(a).

The conclusion that stars of mass higher than about $1.35M_\odot$ are the product of an accretion process acting on stars produced originally with a lower mass is supported by this analysis — despite the imperfections of the model.

5.9 STELLAR ROTATION

We began our discussion of star formation by considering the rotation of stars. Early theorists were concerned with the problem of why stars rotate so slowly and actual observations of stellar equatorial speeds, as illustrated in Fig. 5.1, suggested that stars of spectral type F5 had some special significance.

The formation of stars on the lower branch of star formation is by the collision of two turbulent elements, which are streams of gas in the cloud. In Fig. 5.20(a) we see the effect of the head-on collision of two streams of gas with similar densities and speeds. It is readily seen that there is no systematic tendency for the compressed region to gain angular momentum. In Fig. 5.20(b) the two streams are now offset so that there is some angular momentum associated with the system. Here it will be seen that the compressed region will be rotation-free and that the angular momentum will be associated with the peripheral uncompressed material which would not become part of any resultant star. Finally there is depicted in Fig. 5.20(c) the oblique collision of two streams. If this is illustrated referred to a fixed centre of mass of the system, as in Fig. 5.20(d), then it is seen as the collision of two streams with an offset of centres, and different speeds and densities, but the conclusion is unchanged that there would be no systematic tendency for the compressed region to possess angular momentum.

The conclusion just reached depends on the uniformity of the streams across their interacting cross-sections, or at least in the central region where the star is actually formed. While it seems intuitively reasonable that the central region of a stream might be in a fairly uniform motion, the degree to which this is so cannot be quantified. However, there is another way in which a newly formed star is constrained to rotate slowly. The new star has a density which is only a few times greater than that of the surrounding medium and so it will be strongly coupled to that medium. This coupling is mediated by an exchange of material with both accretion and abrasion taking place. Woolfson has shown that this coupling persists until the radius of the star falls to a particular value.

$$r_P = 2GM/V^2 \tag{5.6}$$

where V is the relative speed of the star and medium and M the mass of the star; thereafter the star will collapse with conservation of angular momentum. For a solar-mass star this will give an equatorial speed of 8 km s^{-1}; a low value, although not as low as the observed speed of 2 km s^{-1} for the sun.

There are other ways in which a young star might lose angular momentum; for example, by the shedding of material during a possible T-Tauri stage of its evolution. The lost material would be in the form of a plasma and due to its charge it would strongly interact with the magnetic field of the star. Out to a certain distance the

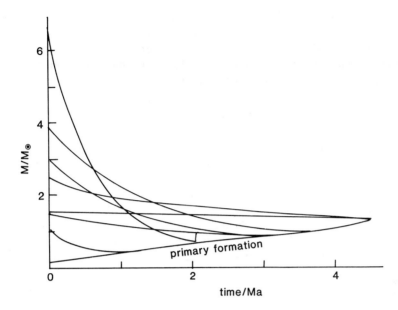

Fig. 5.19 — Accretion lines for stars formed on the lower branch moving through different density regions at different speeds.

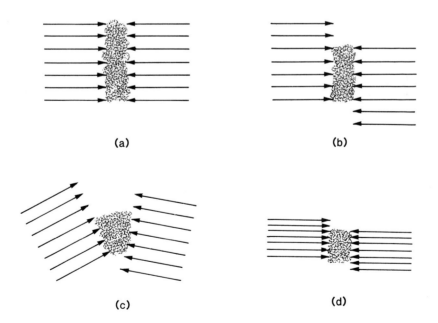

Fig. 5.20 — The collision of two streams. (a) Head-on. (b) Parallel but offset. (c) In general directions with respect to each other. (d) As (c) but relative to centre of mass.

motion would then be isorotational, which means that as the material moved outwards it maintained a constant angular speed around the star. This effect could explain the loss of a few (less than ten) kilometres per second of equatorial speed.

Another mechanism, that suggested by Hoyle and mentioned in section 5.1, involves the magnetic coupling of a collapsing star to external material. Hoyle postulated that the coupling would persist to high stellar density, but this is because he took an unrealistically high value of the stellar magnetic field. Observation of the polarization of hydroxyl maser sources from star-forming regions suggests fields of order 10^2 nT, well below Hoyle's requirement. However, magnetic coupling with an external medium could possibly be effective during the early stages of stellar collapse while the protostar density was still fairly low.

Having established that a star which had formed on the lower branch would have little rotation, we must now consider the effect of accretion. Since material is now joining the star from considerable distances, any assumptions about the uniformity of the medium must be modified. Woolfson has estimated the expected angular momentum of a star which has grown by accretion, and in Fig. 5.21 these results are

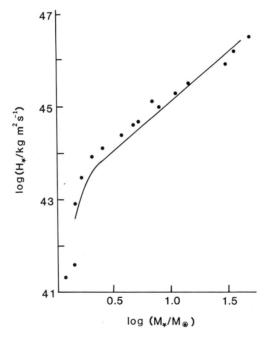

Fig. 5.21 — The observed relationship between the angular momentum of a star and its mass (full line). The theoretical points are derived from the model which gives Fig. 5.15.

compared with observed angular momentum as deduced from stellar observation and theoretical density distributions within stars. The agreement is astonishingly good, but of more importance than the detailed numerical agreement is the agreement of the general form of observation and calculation, including the distinctive droop in the curve for lower masses. For the straight region the observed relationship, for stars earlier than spectral class F5, is of the form

$$\text{Angular momentum} \propto (\text{mass})^{2.1} \tag{5.7}$$

which seems not to have been previously noted. From the theory developed by Woolfson it is found for the more massive stars that

$$\text{Angular momentum} \propto (\text{mass})^{2} \tag{5.8}$$

which agrees quite well with observation.

5.10   GENERAL COMMENTS ON THE STAR-FORMATION MODEL

For a model of such simplicity the agreement of theoretical results with observation is surprisingly good. The starting point of the model, a turbulent cloud, is one that may be readily accepted since it is consistent with observations of water and hydroxyl maser emission regions where stars are believed to be forming. The regions from which the emissions come, of overall dimensions 10^{11}–10^{13} km, are of the same order as the size of the model clouds when stars are forming. Individual sources within the star-forming regions have dimensions 10^{8}–10^{9} km, consistent with the diameters of protostars formed by the model. Again, observed Doppler-shift velocities of order 20 km s^{-1} recorded from different part of the emitting region are comparable with the turbulent speeds given by the model during the later period of star formation.

A pleasing feature of the model is the significance of F5 stars in both the mass–age correlation and also in equatorial velocity, expressed as a distinction between stars which have and have not experienced appreciable accretion. The agreement between observation and theory in relating mass and angular momentum has already been pointed out, but another interesting feature is that the model predicts completely random directions for stellar rotation axes. A difficulty with many previous models is that they derive the stellar rotation too directly from the rotation of material of a nebula, which would give the individual stars highly correlated rotations.

The model has been tested with a wide range of initial parameters — temperature, cloud density and initial turbulent energy. The critical role of stars with mass about 1.3–1.4$M_\odot$ and the prediction of a mass index of about -2.5 are stable characteristics over a wide range of initial parameters. This is particularly pleasing; models which can only match observations by the most fastidious choice of parameters ought to be suspect, especially when the observations to be explained have comparatively little variation from one stellar system to another.

6

The basic capture mechanism

6.1 INTRODUCTION

It is fair to say that by the early nineteen-sixties the nebula theories of the origin of the solar system had regained their former popularity. The often-quoted objections to the tidal theory of Jeans (section 1.3) had gained almost universal acceptance; indeed these seem to have become rather better known than the theory itself. Consequently it is most instructive to re-examine the ideas of Jeans as expounded in his classic work *Problems of cosmogony and stellar dynamics* published in 1919.

Jeans was clearly well aware of the requirements of any theory relating to the scale of the solar system — in particular that the radius of Neptune's orbit is 6400 times the solar radius, so that to draw material out from the sun by tidal action with sufficient angular momentum to form Neptune presents a major problem at the very least. Russell's objection has been quoted in Chapter 1 and this makes it abundantly clear that the present-day planets cannot have arisen from the sun with its present radius. Can we really believe that Jeans did not appreciate this problem? In fact an examination of his own presentation of the theory shows us that he assumed the radius of the sun to be about 30 AU at the time of the tidal event. This feature seems to have been completely ignored by the detractors of the tidal theory. Although the problems of orbits intersecting the solar body, and of the sun being given too much angular momentum (section 4.1) remain, a re-examination of the theory in the light of modern ideas of stellar evolution would be interesting.

To return to the modern nebula theories, it must be reiterated that the problem of angular momentum distribution in the solar system is a major stumbling block. If the planetary material was shed by a contracting protosun, then the solar spin ought to be much greater than is observed. As we have seen in section 3.5, various schemes for the transfer of angular momentum from the protosun to the planet-forming material have been developed. These have been reviewed by Woolfson (1984) who concluded that none of these schemes can be successful.

Since the Second World War a few researchers have concluded that the planetary angular momentum, being so much greater than that of the sun, must be derived from an external source. This idea implies the capture of planetary material, a process which was poorly understood in the context of classical celestial mechanics. Nevertheless the process of capture was examined by Schmidt in the USSR and Lyttleton in England (section 3.2). These astonomers proposed theories of planetary formation essentially independent of stellar evolution and as a consequence were able to solve the angular momentum problem.

More recently, in 1964, Woolfson proposed the Capture Theory which will be the main subject of this chapter. The Capture Theory *is* connected with stellar evolution and is dependent on the formation of stars in large clusters, the existence of which is not in doubt. Before describing this theory in detail we re-examine some aspects of the work of Schmidt and Lyttleton.

6.2   CAPTURE AND ACCRETION

In his 1947 theory, Schmidt proposed that planets were formed from the material of an interstellar cloud captured by the sun in the presence of a third body, another star. Although from energy considerations there is no reason why capture cannot occur when three bodies are involved, the process had not been demonstrated prior to Schmidt's work.

A major difficulty in analysing the motions of three gravitating bodies is that, in general, there is no full analytical solution to the problem. The problem of analysing the motion of an isolated system of two bodies moving under mutual gravitational forces was solved by Newton in the seventeenth century. The relative motion of the bodies satisfies a conic section (ellipse, parabola or hyperbola), the type being dependent on the energy of the system (see Fig. 6.1). Provided initial positions and velocities at some specified time are known, then the relative position of the bodies may be easily discovered at any time in the future (or the past). The solution does depend on the bodies behaving as point masses; a property which is possessed by spherical bodies. Planets are of course very nearly perfect spheres.

When we wish to study the motions of systems containing more than two bodies, there is no difficulty in obtaining the forces which they exert on each other since these satisfy the inverse-square law discovered by Newton, i.e. the force of attraction between any two point masses is proportional to the product of their masses and inversely proportional to the square of their distance apart. Expressed mathematically this appears as

$$\text{Force} = \frac{GM_1M_2}{d^2} , \tag{6.1}$$

where M_1 and M_2 are the masses and d is their separation (Fig. 6.2). Thus we can construct a mathematical model which will be satisfied by our system of three bodies. Unfortunately it is not possible to provide a complete picture of the motions of the bodies by purely analytical techniques. Given initial conditions, only the energy and angular momentum of the system may be obtained easily (of course these are very important because they are conserved and hence constant for all configurations of the bodies). There are some special techniques which can be used to obtain solutions for the relative positions of the bodies at different times. These methods come under

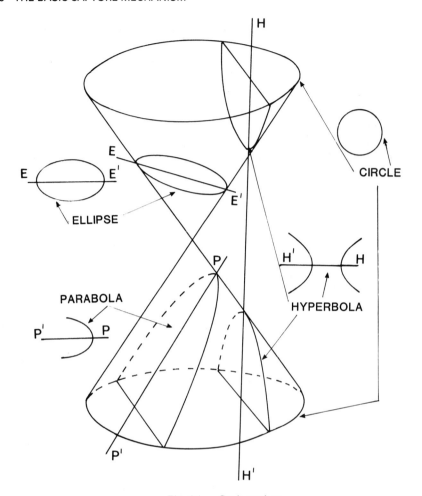

Fig. 6.1 — Conic sections.

Attractive force on each body is

$$\frac{GM_1 M_2}{d^2}$$

where G is the constant of gravitation.

Fig. 6.2 — Gravitational attraction between two bodies.

the heading of numerical analysis and are very commonly used with the aid of computers at the present time. A detailed description of such methods is beyond the scope of this work, but the basic idea involved is that of proceeding to calculate solutions in very small time steps. If we compute many such steps then we may have obtained the positions of the bodies over a significant amount of time. There are going to be errors in such a calculation, but these can be reduced to an acceptable magnitude by careful choice of time steps (see Appendix).

Numerical methods are ideally suited to computers, which can do arithmetic extremely quickly, but in 1944, when Schmidt tackled his theory, computing aids tended to be of the handle-turning mechanical variety.

Since the mathematical model for any purely gravitational interaction permits the direction of time to be reversed, Schmidt found it easier to consider the removal of a body orbiting the sun by the perturbing action of a passing star, rather than to try to guess under what conditions capture might be possible. The capture process can be considered as the 'mirror image' in time of the orbital disruption. Thus the supposed interstellar cloud was taken to be in elliptical orbit about the sun at the start of the calculation, and the passing star, moving on a heliocentric hyperbolic path was 'aimed' to make a close passage through the system. Schmidt found that it was possible to perturb the cloud (a portion of which is represented by a point mass) so that it escaped from the sun. Viewed in reverse, the process is equivalent to a cloud being captured by the sun. A real interstellar cloud would behave in a more complicated fashion, since it is an extended body subject to differential gravitational forces, but it is clear that such a triple encounter could result in some material being captured by the sun. Any planetary bodies forming from the cloud material could have appropriate angular momentum since this would be derived from the original relative orbit. An event as envisaged by Schmidt is depicted in Fig. 6.3.

The above theory might be criticized on the basis of being too unlikely since encounters between galactic field stars must be very rare, and the event considered here involves a gaseous cloud in addition to the stellar bodies. Nevertheless we do not have strong evidence of other planetary systems and we cannot rule out completely the possibility that our system is unique; an idea that Jeans seems to have found acceptable.

An alternative to the three-body process was proposed by R. A. Lyttleton in 1961. In this theory it was proposed that material would be captured by the sun as it passed through an interstellar cloud of gas and dust. Particles of the cloud moving past the sun would be deflected by the gravitational attraction and hence 'focussed' on to an accretion line in the direction opposite to that of the sun's relative motion (Fig. 3.3). In the region of enhanced density near to the accretion line many inelastic collisions would occur, destroying much of the kinetic energy of the particles involved. As a result, a substantial amount of material could be captured and enter orbits bound to the sun. We shall not give further details of this accretion theory here since it has already been discussed in Chapter 3.

6.3 THE CAPTURE THEORY

Both of the above 'capture' theories of the origin of the solar system still rely on random processes occurring in a dispersed medium for the actual production of

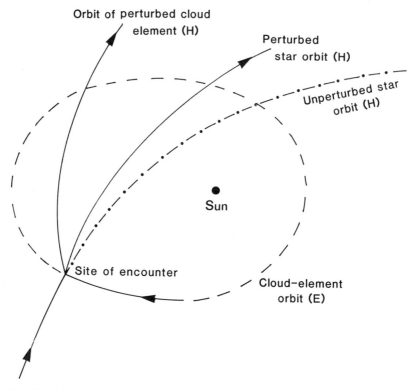

Fig. 6.3 — Schmidt's encounter model. The diagram shows the disruption of a sun–cloud binary *or* (with time reversed) cloud capture by the sun in the presence of a third body. H and E indicate hyperbolic and elliptical orbits respectively.

planets. A satisfactory description of such processes is still to be found, despite the close attentions of many nebula cosmogonists (e.g. Safronov, 1972). The only *undisputed* mechanism permitting the condensation of tenuous material to form relatively dense bodies is gravitational instability, and all theories of star formation rely on this process. Consequently it seems reasonable to propose that planets are formed in a similar way, and this is precisely what Jeans suggested in his tidal theory. Of course a cloud of planetary mass mainly consisting of gaseous material will not condense if it is too hot; the thermal energy must be sufficiently low if gravitational forces are to dominate.

Bearing in mind the reasons for the failure of the tidal theory of Jeans and also the partial success of Schmidt's hypothesis, it is logical to propose two major requirements of a theory of the origin of planetary systems. These are:

(1) planetary material must be captured (to solve the angular momentum problem);
(2) primitive planetary material must be sufficiently cool and dense to be gravitationally unstable.

The Capture Theory, proposed by Woolfson, satisfies both of these requirements. In this theory it is proposed that an encounter took place between the sun and a cool protostar, both bodies having originated in the same stellar cluster. From the results described in Chapter 5 it turns out that the density of stars in a young cluster would be sufficiently large to make close encounters fairly common. Being somewhat less massive than the sun, the protostar would be at an earlier stage of development and therefore be of low density and correspondingly low temperature. During an encounter in which the closest approach distance was not much greater than the physical dimension of the protostar, it would be greatly affected by solar tidal forces. Material would be removed from the protostar and Woolfson proposed that some of this was captured by the sun and eventually formed the planetary bodies. In some respects this theory is similar to that of Jeans, but it has the great advantage of being able to satisfy the angular momentum requirements much more easily. Also the material from the protostar would be relatively cool, thus making the prospect of planetary condensation much more likely while not involving the condensing planet in a possible collision with the sun.

Of course the idea of capture when only two bodies are involved may be a little unconvincing since it is well-known that the relative motion of two bodies takes the Keplerian form. However, it must be stressed that the Capture Theory proposes that the protostar is a very extended body subject to tidal modification; the behaviour of such a body is not well-described by that of a single point mass.

The analysis of the dynamics of massive extended bodies is not possible by classical methods. Fortunately the Capture Theory hypothesis coincided with the development of the first computers capable of being used for numerical testing of such models. In the nineteen fifties the available machines would have been inadequate for the massive amounts of arithmetic operations necessary for this type of work. As a consequence Woolfson was able to test his new theory in a more comprehensive manner than any previous cosmogonist. The results indicated the plausibility of the capture hypothesis beyond any reasonable doubt. A schematic view of the proposed capture event is shown in Fig. 6.4. The rest of this chapter will be devoted to a more detailed description of the Capture Theory, although it will be appreciated that the modelling involves mathematical techniques well beyond the scope of this work.

6.4   TIDAL EFFECTS

Most people are aware of the existence of tides but perhaps not many would connect the rise and fall of sea-levels with the origin of the solar system. Nevertheless, when we mention tidal effects in this work we are referring to precisely the same mechanisms which cause the water to flow in and out round the shores of open seas. The forces responsible for tides on the earth are due mainly to the moon and, to a lesser extent, the sun.

The tidal effect of the moon on the earth is illustrated in Fig. 6.5. The side of the earth which faces the moon is subject to a larger gravitational force (Fig. 6.2) than the centre of the earth, simply because it is nearer. In turn the opposite side of the earth is further away than the centre and thus experiences a lesser force. These differences in the gravitational forces (differential forces) cause the earth to be slightly distorted, resulting in tidal 'bulges' being formed along the direction of the line joining the

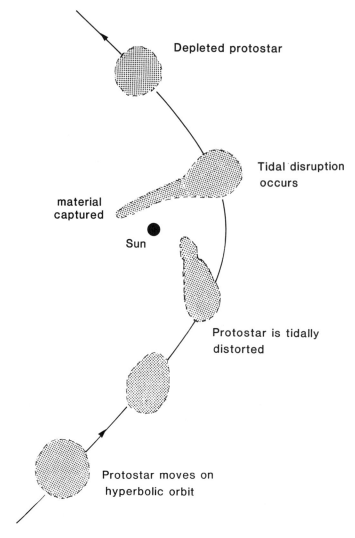

Fig. 6.4 — Capture theory model. A schematic view of the sequence of events in the Capture Theory.

centres of the two bodies. At least this would be the case if the earth did not rotate. Since the material of the earth cannot react instantaneously to changing forces it finds itself carried away from the moon–earth line before it can achieve maximum distortion. The angular discrepancy is quite small, being only about two or three degrees. Changes in sea level due to tides are much greater than solid-body distortion because water is much less viscous than the terrestrial crust and has no mechanical strength. The amount of distortion in the earth is very small in relation to its size. Even when magnified by the effects of irregular coastlines, tide heights rarely exceed

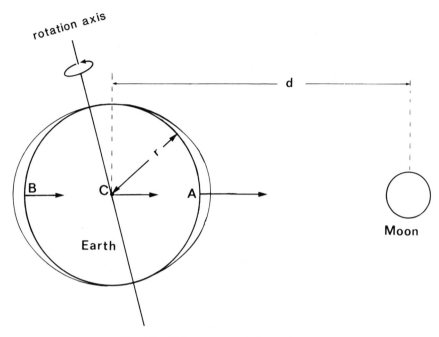

Fig. 6.5 — Tides on the earth due to the moon.

10 m. The dual bulges ensure that a maximum tide height is reached twice per day, as the earth spins under the bulges which are directed approximately towards the moon with its much slower orbital motion. Since the moon does not move in the equatorial plane the diurnal maxima are of different magnitude.

A side-effect of this continuing tidal activity is the slowing of the earth's rotation caused by friction between the spinning earth and the tidal material. Consequently conservation of angular momentum in the earth–moon system means that the moon must be moving away from the earth. Eventually the earth will rotate in exactly the same period taken by the moon to complete an orbit and then the tides will be 'frozen', in the same way as are those of the moon. The tidal effect of the sun, which is by no means negligible, has been omitted from the above discussion for the sake of clarity.

Even though the moon is our nearest celestial neighbour, its tidal effects are small because its mass is small. Consider the sequence of events which would occur as a result of two massive bodies (stars or planets) coming very close to each other (see Fig. 1.4). Obviously the distortion due to differential forces will increase as separation diminishes. If the bodies are very close, the tidal forces on the facing sides will exceed those acting on the opposite sides because of the inverse-square law. In these circumstances the bulge on the near side will be larger than on the other, resulting in a 'pear'-shaped configuration. If we continue to reduce the separation, our distorted body may start to break up, with material spilling out from the pointed end; this effect is known as tidal disruption. It is a relatively straightforward matter to calculate the critical parameters associated with tidal disruption of this form and these are shown graphically in Fig. 6.6. We note that the critical separation at which

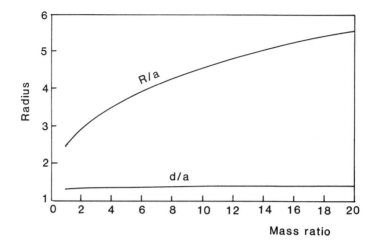

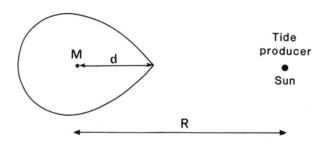

Fig. 6.6 — Critical tidal parameters. variation of critical distortion and the distance for disruption with the mass ratio (S/M). The undistorted radius is a.

disruption commences increases (in terms of the undisturbed radius of the disrupted body) as the mass ratio increases. In contrast the shape of the body immediately prior to disruption does not vary greatly.

The above results apply directly to one-dimensional relative motion. A protostar approaching the sun would move on an orbit classified as hyperbolic and so we must extend our analysis to two dimensions. Also we must be prepared to consider the dynamical behaviour of tidal growth; the results above apply to the equilibrium situation in which it is assumed that the body shape responds perfectly and instantly to the changing gravitational field. In reality no body could achieve such a response since shape changes must be due to internal motions of finite velocity.

As a protostar approaches the sun, the direction of the largest gravitational field gradient (along which the greatest distortion should occur) rotates in the same sense as the orbit itself. For reasons already given, the tidal tip of the protostar cannot 'keep up' with this rotation whose rate is increasing as perihelion approaches. Thus the tip will always point in a direction as decreed by the field gradient at an earlier instant. The size of this lag will depend on the actual orbit and also on the primitive rotation of the protostar. After perihelion the tidal lag will eventually change into a tidal lead because the orbital rate is now decreasing. This process is illustrated in Fig. 6.7. These results have been computed under the assumption that the distortion would remain constant; this would not be true in practice because, as we have seen earlier, the tidal distortion would grow as the protostar approaches perihelion. However, the calculation is useful from the point of view of capture of planetary material. Due to the rotation, material near the tidal tip is moving more slowly relative to the sun than the centre of mass of the protostar at perihelion. If it can be treated independently this implies that such material may have insufficient energy to escape from the sun, which means that it will be captured into a heliocentric elliptic orbit. It will be clear that the tidal disruption of the protostar provides the opportunity for capture.

6.5   SIMPLE MODELS OF DISRUPTION AND CAPTURE

Any new theory of cosmogony begins in an intuitive way. The cosmogonist, having an extensive knowledge of physics and celestial mechanics say, is able to generate a hypothesis which might explain the origin of some astronomical system. Such a hypothesis can be plausible only if it is seen to be consistent with observation and accepted physical laws. Consequently it is necessary to examine the theory in a quantitative sense, usually involving mathematical techniques. Typically, a mathematical 'model' is constructed. This will consist of some system of equations connecting the various physical quantities (variables) which affect the hypothesis. Details of 'models' used in capture theory research are beyond the scope of this work, but some idea of what is involved can be obtained by reference to the Appendix, which contains a 'model' of one-dimensional gravitational motion. Fortunately it is possible to gain some insight into the testing of the Capture Theory without involving mathematics.

Naturally it is desirable to start with a discussion of the most basic aspects of testing. To some extent we have already done this in the previous section on tidal effects, but in this section we shall be more quantitative.

The tidal disruption of a body (e.g. a protostar) has been described qualitatively above. We now pose the question: how close must the bodies approach before tidal disruption occurs? This was the topic of the analysis in section 3.1, leading to equations (3.1) to (3.2). Roche (1854) showed that a satellite in circular orbit about a planet would be tidally disrupted if the radius of its orbit was less than 2.45 times the planetary radius. This assumes that the planet and satellite have the same density; if the satellite was less dense then the critical distance is increased. The existence of Saturn's rings has been quoted as evidence of the tidal break-up of a satellite with a decaying orbit. The Roche 'limit' is not directly applicable to the disruption of our protostar since it would have a hyperbolic orbit relative to the sun. This means that

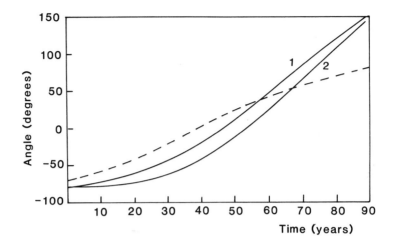

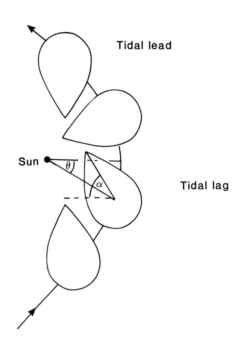

Fig. 6.7 — Tide rotation. Rotation of the tidal tip of a distorted body during an encounter.
Angle θ — — —; Angle α ($\dot{\alpha}_0 = 0.05\theta_0$), $\frac{1}{\rule{2em}{0.4pt}}$; Angle α ($\dot{\alpha}_0 = 0.40\theta_0$), $\frac{2}{\rule{2em}{0.4pt}}$.

the tidal forces acting on the protostar would not persist as in the Roche problem,
and the limiting distance would be smaller; Jeans showed that this was the case. Fig.
6.8 shows the critical separation for tidal disruption in two cases of eccentricity of the
relative orbit. It will be seen that the minimum approach distance is reduced for the
higher eccentricity.

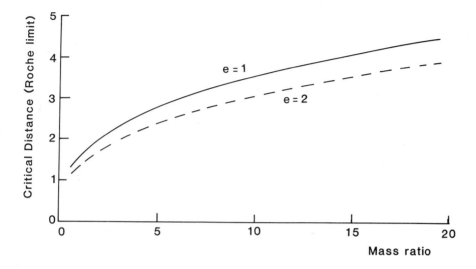

Fig. 6.8 — Roche limit. Variation of the critical separation for tidal disruption with mass ratio. Two relative orbits of different eccentricities are considered. The critical distance is smaller for the greater eccentricity.

An additional factor to be assessed from the Capture Theory point of view is the evolutionary state of the protostar. According to theories of star formation (see Chapter 5) a protostar with mean density approximately 10^{-8} kg m^{-3}, such as we are proposing, would not be in equilibrium; it would be contracting towards the main sequence. The actual rate of collapse is uncertain, but in our 1971 paper we assumed that Kelvin–Helmholtz contraction would be realistic (our views in this respect have changed more recently, as will be clear later in this chapter). The Kelvin–Helmholtz contraction is governed by the rate at which the body is able to radiate energy, assuming a constant effective temperature and opaque conditions. For a protostar of one-quarter solar mass, radius 16.7 AU and temperature 50 K, the collapse rate would be 0.2 km s^{-1}. A more realistic value would be in excess of this, but the upper limit is the 'free-fall' value of about 5 km s^{-1} which would be attained by material having fallen from infinity.

To investigate the critical parameters for tidal disruption and capture, a fairly simple two-dimensional computer simulation was constructed. A three-body gravitational system comprising point masses representing the sun, protostar, and a small particle resting initially on the contracting surface of the protostar was used to form the model. Since the relative motion of the sun and protostar is assumed to be hyperbolic this is easily computed by classical means. The motion of the small particle can be computed by numerical methods. Many encounters modelled in this way were simulated and Fig. 6.9 provides an illustration of the results obtained. The simulation yielded critical distances for disruption which were very similar to results obtainable from Jeans' analysis. It was seen that as the perihelion of the protostar orbit increases, the major axis of the orbit of the captured particle rotates in a clockwise direction. For a case in which the protostar just enters the critical region

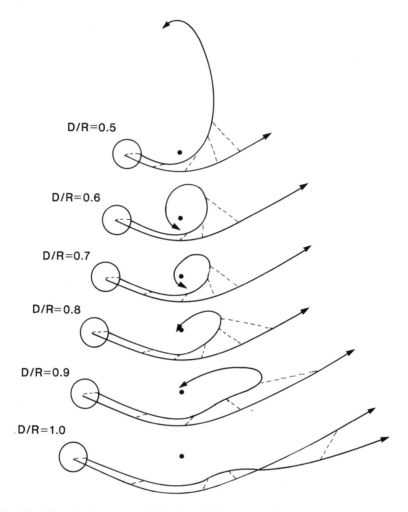

Fig. 6.9 — Three-body capture orbits in hyperbolic encounter (eccentricity = 2, S/M = 4, star radius = 16.7 AU). D/R is the ratio of the perihelion distance to the Roche limiting value. Dashed lines indicate simultaneous positions of the centre of the protostar and the particle.

the particle is captured effectively at the aphelion of its initial orbit. If the protostar perihelion is half the critical value the particle orbit has its aphelion on the opposite side of the sun.

These results are very significant from the point of view of the theory. A close encounter is seen to produce initial orbits on which captured material recedes rapidly from the sun immediately following capture. If this were not the case, it is difficult to see how material torn from the protostar could condense under its own gravitational field while being subject to growing tidal forces. It should be emphasized that the three-body simulation gives only a crude picture of the capture encounter. The protostar would be an extended body which, when tidally distorted, would certainly not behave as a gravitational point mass. Nor would material be captured only from near the surfaces of the protostar. It is possible that material from deeper within the protostar would take up initial orbits in the range generated by our variation of perihelion. Clearly we need to perform a full three-dimensional simulation of the encounter in which the detailed behaviour of the protostar can be incorporated. While it is possible to construct mathematical models reflecting a great deal of detail, their analysis is far from easy. It is fortunate that powerful computers, available today, permit us to apply numerical methods for very sophisticated simulations.

6.6   MORE SOPHISTICATED MODELS OF THE CAPTURE EVENT

In the early nineteen-sixties the availability of the new Atlas computer, which was much more powerful than earlier machines, made it feasible for the Capture Theory to be examined in a more detailed manner. To simulate the fluid nature of a protostar whose mass was not wholly concentrated at the centre (which had been assumed previously), Woolfson introduced the point-mass model using two space dimensions. Since a protostar would have some degree of central condensation, most of its mass was concentrated at the centre while the remainder was distributed on a network of points filling the 'area' of the model star (Fig. 6.10). This scheme allows more

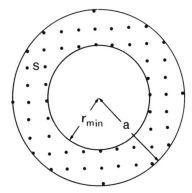

Fig. 6.10 — A model star. Mass points are contained in an annulus with inner radius r_{min} and outer radius a. Most mass is concentrated at the centre.

realistic modelling of the gravitational field of a protostar when it is distorted by tidal forces. Also it makes possible the estimation of the amount of material captured by the sun during an encounter together with the range of initial orbits. In fact it was not necessary to fill entirely the area of the star since the central regions would not be likely to suffer excessive tidal modification. Thus the network of mass points was situated in an appropriate annulus.

   To simulate the capture event using a computer it is necessary to calculate the trajectories of all the point masses, each of which is affected gravitationally and by pressure forces due to all the other points. Thus at each stage of the calculation the number of forces to compute is proportional to N^2, where N is the total number of mass points; hence the need for a powerful machine. The model being described here contained 58 points and so it demanded about 400 times more arithmetic operations per time step than the three body models described previously.

As implied above, it is vital to consider pressure forces as well as those due to gravity in any point-mass model. In this model the pressure forces were of two types. First a radial force was included to prevent an isolated protostar model from free-fall collapse. Second, an inter-point force was specified to prevent unnatural 'pooling' of points which would occur otherwise due to gravitational attaction. This repulsive force included an empirical constant called an 'elbow' factor, a plausible value of which had to be deduced by testing the disruption properties of the model. Such testing of models of this type is essential if we are going to place any confidence in the simulation of the capture event. The means of testing is based on our knowledge of the quasi-static behaviour of an extended body under tidal forces. By varying the empirical parameters for a model star subject to static tidal forces, it is possible to determine the most suitable values. The results of simulating the capture event by using the above scheme are illustrated by Fig. 6.11. Here a sequence of configurations of the model star are plotted, culminating in a stage just after perihelion when it is clear that a number of mass ponts have been removed from the protostar. To determine the orbits of these mass points they were treated separately as a sequence of three-body problems. Their trajectories were computed under the influence of the sun and the receding protostar until their energies with reference to the sun became stabilized, indicating that the protostar had negligible effect. Most of the points were found to have been captured by the sun, although a few possessed hyperbolic orbits. The minimum amd maximum perihelia of those in elliptic orbits were found to be 1.4 AU and 38.4 AU; values within the range observed in the present-day solar system. It must not be assumed that the mass points represent individual planets; rather that the orbits obtained demonstrate the range of achievable values. The effective mass of each point was about 50 earth masses, and the planets themselves could be thought of as arising from the partial gravitational condensation of groups of points. Planets formed from this material would move on orbits far more eccentric than those currently observed in the solar system (excepting comets). The reduction of orbital eccentricities will be considered in Chapter 9.

The detailed model (published in 1964) confirms in a more convincing manner many of the results inferred from the more simple simulations. It also serves as the prototype of many subsequent models, highlighting the advantages and disadvantages of mass-point simulation for continuous fluid systems. A big advantage of mass points is concerned with the simplicity of model construction, it is a straightforward matter to form the necessary mathematical equations of motion which need to be solved. Also these equations appear, at first sight, to be amenable to solution by reliable methods of numerical analysis with computer assistance. Of course there is the disadvantage that the mass is not distributed in a continuous manner, as it is in stellar bodies. The occurrence of near-singularities produced by close approaches between mass points must be avoided, and this can be achieved by the introduction of repulsive forces ('elbow' factor). A 'real' protostar would be three-dimensional, unlike the plane model constructed in 1964. In principle a three-dimensional model can be produced just as easily, but the problem of resolution is increased when points

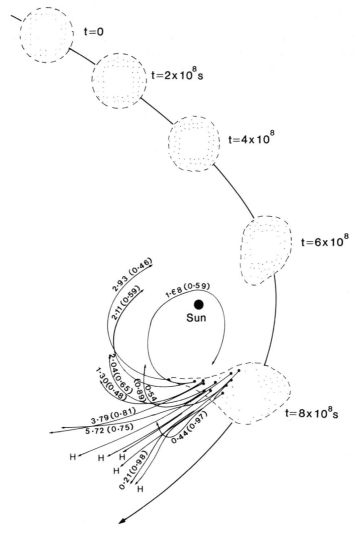

Fig. 6.11 — The loss and capture of material from a model star. Figures marked on orbits are perihelion distance (10^{12} m) and, in brackets, eccentricity. H signifies a hyperbolic orbit and represents material not captured. Initial star parameters:
mass = $0.15\,M_\odot$, radius = 2.2×10^{12} m, perihelion = 6.67×10^{12} m. (From Woolfson, 1964.)

are distributed within a sphere. This can be seen if we count the points distributed on a cubical lattice contained within a sphere, with lattice spacing half the radius we obtain 33 points, and with a spacing one-third of the radius 123 points are needed. In the two-dimensional case the number of points 'filling' a circular disk are respectively 13 and 29. Thus if we wish to maintain similar space resolution in a three-dimensional simulation the computational complexity (being proportional to the square of the number of points) is greatly increased. In practice the computer processing time requirements increase even more sharply than is suggested by the above figures. It

will not be too surprising that much effort has been channelled into the development
of models in which the above problems are alleviated. Consequently the first three-
dimensional star-encounter simulation (Dormand and Woolfson, 1971) avoided
completely the direct determination of inter-point forces. It was noted that the
profile of a tidally distorted body would have a shape similar to that of part of a
limaçon curve (Fig. 6.12) which has the equation in polar coordinates

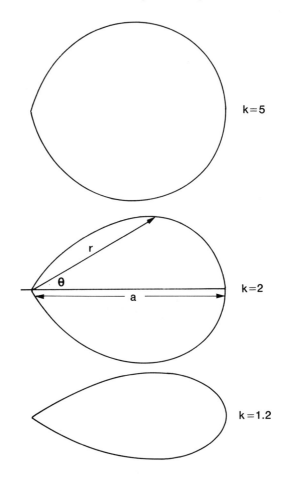

Fig. 6.12 — The limaçon $r = a(k \cos \theta - 1)/(k - 1)$, $\cos \theta \leqslant 1/k$, for three values of k.

$$r = \frac{a(k \cos \theta - 1)}{(k - 1)} \, ,$$

$$k > 1 \, , \qquad \cos \theta \leqslant \frac{1}{k} \, ,$$

where a is the length of the axis, and k is the 'eccentricity' of the limaçon. Values of k

near to unity yield very elongated figures and the limaçon tends to a circle as $k \rightarrow \infty$. Rotation of the limiçon about its axis forms a surface appropriate to our requirements. A body bounded by such a surface is termed a limacoid. In our star-encounter model a limacoid was 'fitted' to the configuration of mass points at each integration step and this was used to produce a gravitational field which determined the motion of the points during the next step. By the word 'fitted' we mean that the appropriate values of k and a were determined to match as well as possible the volume containing all, or nearly all the mass points. It will be clear that during the early stage of the computation large values of k would result, indicating slight tidal distortion. As the simulation proceeds the k value decreases corresponding to more severe distortion. Fig. 6.13 illustrates the limacoids fitted to a model star just prior to

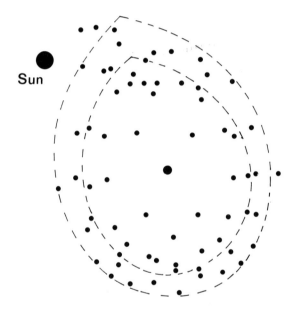

Fig. 6.13 — Configurations of star mass points with fitted limacoids. Outer shell: $k = 2.27$, $a = 37.2$ AU; inner shell: $k = 2.44$, $a = 28.3$ AU.

perihelion passage. In this case the model star comprised initially two spherical concentric shells of mass points distributed at the vertices and on surface normals of regular polyhedra, and so the gravitational field calculation was based on two limacoids. Three-quarters of the total stellar mass was concentrated at its centre; this assumption is reasonable since we do not expect the inner regions of the star to be affected greatly by the encounter.

In addition to the gravitational forces, the mass points were subject to a central repulsive force in order to prevent collapse at a rate exceeding the Kelvin–Helmholtz rate (section 6.5). Any points outside the limacoid were treated separately and the 'pooling' of these points was prevented by a process of velocity-sharing, whereby after each integration time step each point was reallocated a velocity including a component from neighbouring points. This introduced into the model the character-istics of a fluid with viscosity.

The main advantage of the limacoid model over those considered previously is the absence of most inter-point forces since these are contained in the surface-fitting procedure. Consequently the model is much easier for numerical analysis. Increasing the number of points would not be as expensive as for pure mass-point schemes. However, the model does assume some symmetry, which will be destroyed by the event under consideration. For this reason it is feasible only to continue such a simulation while the star resembles a limacoid. Some extension to the period was obtained by separate treatment of points torn away from the protostar. A result of a limacoid simulation is illustrated by Fig. 6.14 which shows a sequence of configu-

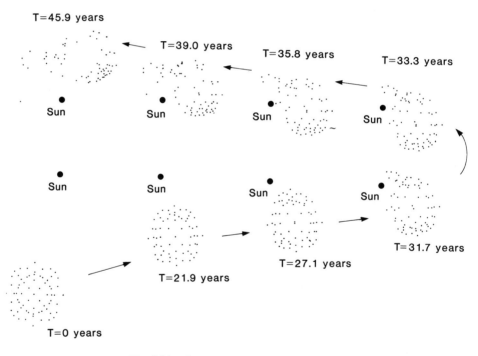

Fig. 6.14 — Star encounter modelled by limacoids.

rations of the model star ending shortly after perihelion. To find the orbits of captured points, further integration was restricted only to those deemed to have been removed from the protostar. This simulation was continued until the orbits had become stable, i.e. they were no longer affected significantly by the gravitational field of the receding star. These orbits are drawn in Fig. 6.15 and their relevant parameters are listed in Table 6.1. The scale of these orbits is similar to that observed

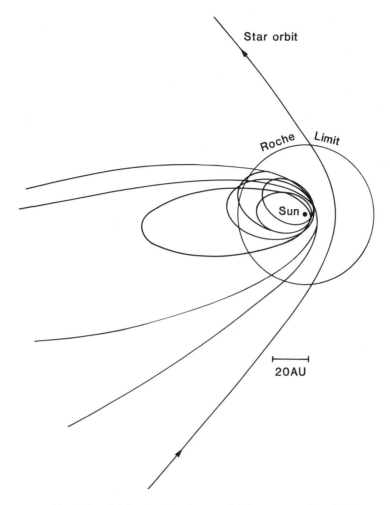

Fig. 6.15 — Orbits of captured mass points in encounter (Fig. 6.14).

for the solar system but, of course, the eccentricities are much larger. We shall see in Chapter 9 that this discrepancy is not an important drawback for the theory. Indeed further studies will suggest that the existence of highly eccentric orbits in the early solar system is an essential prediction for any plausible cosmogonic hypothesis.

In total, four limacoid simulations were completed using different initial conditions. Fig. 6.16 shows the mass distribution obtained in all cases compared with the solar system at the present time. Although none of the simulations matches precisely with the solar system, the distributions bracket the true distribution and are similar to it in form. The presence of a central peak in each distribution is significant and can be related to the rate of ejection of material by the protostar; the nature of the tidal forces suggests very strongly that the star will eject material most rapidly near the perihelion of the orbit. It is likely that a painstaking search would yield initial

Table 6.1 — Orbits of captured mass points

Point	Semi-latus rectum (AU)	Eccentricity	Mass (Jupiter units)
1	16.0	0.938	1
2	15.5	0.979	1.5
3	11.3	0.755	0.5
4	10.5	0.756	3
5	6.3	0.784	1
6	7.7	0.920	1
7	7.7	0.730	1
8	6.5	0.780	2.5
9	0.3	0.995	2

Protostar parameters: Mass $M_\odot/4$, radius 16.7 AU, perihelion 18.6 AU, eccentricity 1.5.

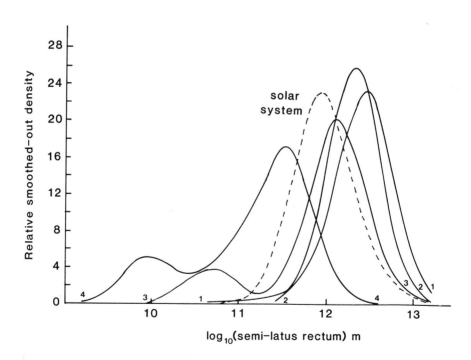

Fig. 6.16 — Comparison of the solar-system mass distribution with that of the captured material from four star-encounter simulations.

conditions for a simulation to reproduce closely the solar system mass distribution. However, at the time of the investigation the necessary computing requirements were prohibitive and it was judged that the standing of the theory would not be significantly enhanced.

The limacoid tests also suggest that the filament of material drawn from the protostar would have density sufficient to allow direct planetary condensation during the first orbit around the sun. It will be seen that the orbits (Fig. 6.15) have aphelia on the side opposite to the protostar perihelion and so the material moves away from the sun following capture. The process of condensation will be examined in the next chapter.

6.7   RECENT DEVELOPMENTS IN MODELLING

Although the models described above must be regarded as effective in the simulation of tidal disruption and capture, it is natural to consider new developments in astrophysical fluid modelling. Clearly the use of superior schemes could lead to a better appreciation of the processes involved in the capture event. One obvious improvement would be the application of more mass points than in the earlier models. This would imply more lengthy computer processing, but the availability of very powerful machines does not rule out this option. A second option would be to improve the physical quality of the models: we consider this to be the best way ahead.

In most modelling schemes a fixed number of points are used to mark the mass distribution of the body (e.g. protostar) being simulated. A major difficulty of many such schemes is the potential singularities inherent in gravitational field calculations. Close approaches between points give large forces (equation (6.1)) and, in the limit (zero separation), an infinite force occurs. To avoid this type of event, pseudoforces must be introduced. For example, repulsive forces are intended to simulate the behaviour of pressure forces in preventing the growth of unrealistically high densities. Since the pressure force depends on the gradient of the density, a 'good' model will permit the estimation of this quantity at any point in the region occupied by the fluid body. Such a model was developed in 1977 by Gingold and Monaghan, who described the procedure as 'smoothed particle hydrodynamics' (SPH). A full appreciation of this scheme requires detailed mathematical knowledge, but the principles involved are quite straightforward.

In SPH our model protostar will again be represented by an assembly of points. However, each of these now will mark the position of a fluid element whose mass is spatially distributed in a spherically symmetric manner. The extent of this distribution is termed the smoothing length and this will be large enough to ensure that fluid elements overlap (Fig 6.17). It is reasonable to picture the SPH model as a mass-point scheme in which each point is smeared out appropriately to give a continuous mass distribution. This continuity is the key to estimation of the density and its gradient at any position. Naturally it is also possible to compute the gravitational field anywhere, and the gravity force at the 'centre' of an element due to itself will be zero (an object placed at the centre of the earth would be weightless). Consequently no singularities can occur, since two points occupying the same position would not affect each other.

The above properties are extremely attractive for any type of astrophysical

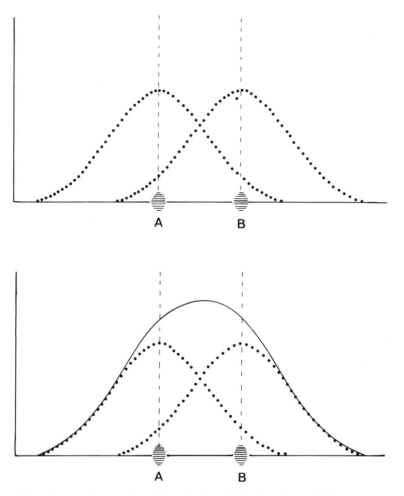

Fig. 6.17 — Separate (top) and combined density distributions for two SPH elements in one
dimension.

modelling of compressible fluids. The physical quantities pressure and temperature
can now be incorporated in a Capture Theory model. If our protostar is very tenuous
we can assume a constant temperature, and then the pressure will simply be
proportional to the density. For parts of the protostar which are more dense we
would expect an adiabatic state defined by

$$p = K\rho^{\gamma} , \tag{6.2}$$

where p is the pressure, ρ is the density and γ, K are constants. The modelling of
viscous effects is also fairly straightforward.

The above description may suggest that all modelling problems are solved by the

SPH scheme. This is far from true. The SPH scheme, in its original form, suffers from a serious drawback when evolving bodies are to be simulated in that it employs a constant smoothing length. This factor limits severely the range of density which can be covered. With the quasi-stable protostar described earlier, large changes in density would not occur during the period of the encounter, but if a protostar were collapsing rapidly, at this time, we would need to improve the model. Current theories of stellar evolution do suggest rapid collapse at the stage in which we are interested. It should be said at this stage that mass-point models also prove difficult for the simulation of bodies in rapid collapse.

A consideration of Fig. 6.17 will give an appreciation of the density limitation. If the two elements centred on A and B approach each other, the greatest density (occurring at the mid-point) will increase. It is obvious that the maximum value will be achieved when the two elements are superimposed (A coincides with B). Thus in a system comprising two elements, the maximum density achievable is bounded by twice the greatest initial density. Since our protostar model will have around 120 elements, a density range of two orders of magnitude could be envisaged if it is gravitationally unstable (collapsing). Assuming homologous collapse this suggests a radius change factor of five. Now we require much larger changes to be simulated, especially in the case of planetary evolution which will be considered later. Consequently we have introduced a variable smoothing length which is dependent on the local density of fluid elements. In the context of Fig. 6.17 this means that the smoothing lengths would shrink as A approaches B, thereby raising the maximum of each distribution. In a centrally condensed protostar model, smoothing lengths near the centre would be smaller than those at the surface. This variation has been found to give a useful extension to the applicability of SPH (Coates, 1980; Allison, 1986; Dormand and Woolfson, 1988).

The validation of any type of mathematical model is an important task. If the model cannot yield good results when applied to test problems, whose solutions are known, there is no reason why we should believe results obtained with regard to anything else. Not surprisingly any test problems are considerably easier than the one at hand; normally they are concerned with spherically symmetric configurations. A fairly obvious test is the simulation of gravitational collapse, for which the true solution is easily found. Any mass-point model is capable of dealing with this situation. A more suitable test for SPH concerns equilibrium configurations for a gravitating gaseous sphere subject to the adiabatic equation of state (6.2). Such configurations are called polytropes and the appropriate density distributions can be obtained exactly by analytical means for some values of γ. Our SPH model performs very well on this test and so we can go ahead confidently with capture simulation.

A fairly large number of encounters between the sun and the SPH protostar have been examined. A particular case is illustrated in Fig. 6.18. The protostar had mass $M_\odot/5$ and initial radius 20 AU. Its initial collapse rate was 25% of free-fall velocity and the relative orbit had eccentricity 1.4 and perihelion 10 AU. About one-fifth of its mass was captured by the sun and Fig. 6.19 shows the distribution of the orbital semi-latera recta of the captured material. As with the earlier limacoid model we observe a pronounced maximum; a feature expected from the contemporary mass distribution of the solar system. Although the sequence of events appears similar to those in previous numerical experiments, there are some differences due to the

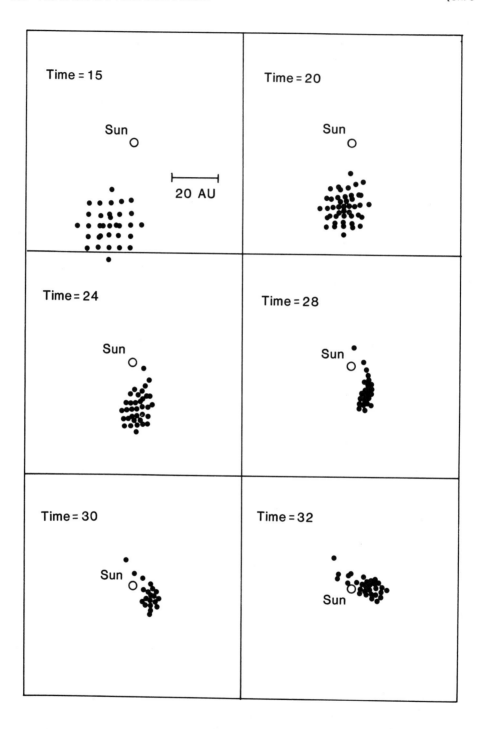

Fig. 6.18 — A sequence of configurations of a tidally disrupted SPH protostar. Times are in years from start of simulation.

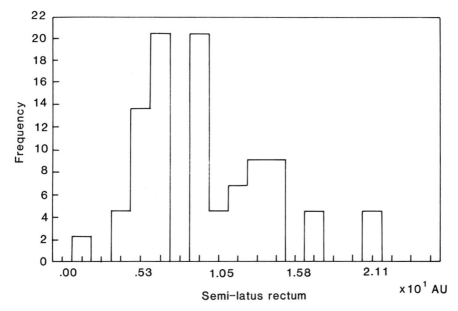

Fig. 6.19 — Distribution of semi-latera recta of orbits of captured elements following encounter from Fig. 6.18.

nature of the initial conditions. The protostar takes on a much more distorted shape well before closest approach, and this can be explained easily by reference to a non-spherical Roche model (section 1.2). Any portion of the boundary more distant than the mean surface from the central mass will have a smaller free-fall velocity than that of the mean. Consequently it will be 'left behind', and so any tidal distortion of a collapsing body will be enhanced by the collapse.

Although the protostar is outside the Roche limit at the start of the simulation, the tidal forces oppose collapse in the direction of the solar radius vector, but not normal to this. Consequently it becomes much more elongated and tends to split rather in the manner for a filament as envisaged by Jeans. This occurs before closest approach. The new results suggest that the density of captured fragments may be higher than was believed previously, thus improving the prospects for planetary condensation. Densities in the protostar and captured material are illustrated in the form of a contour plot (Fig. 6.20). The captured material accounts for about 20% of the initial stellar mass — in other words 25 times the present-day total planetary mass. A small fraction of the material torn from the protostar is not captured by the sun and so will become independent of both stars. We estimate that the protostar will retain about 75% of its original mass.

6.8 CONCLUDING REMARKS

The work described in this chapter confirms definitely the mechanism of tidal disruption and capture which is of central importance to the Capture Theory. The

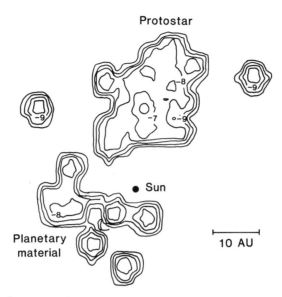

Fig. 6.20 — Density contours in protostar and captured material (from Fig. 6.18). Contour labels are log (density/kg m^{-3}). The outermost contour is − 11.

consequences of a tidal encounter are not sensitively dependent on initial conditions, although we cannot be absolutely sure that these are possible in the context of stellar evolution. It must be said that the type of event being considered, which would lead to a system similar to our own, would not be common according to the theory of star formation of Chapter 5. However, the Capture Theory is not required to predict that planetary formation is an *essential* byproduct of star formation. Also we need not insist that all planetary systems should have the same dimensions as our own.

The first of the requirements of a theory (Chapter 4) has been well-satisfied. There can be no doubt that the distribution of planetary angular momentum is predicted while almost every other theory fails in this respect. The requirements of a cold origin of planets and coplanar direct orbits are also met. The actual formation of planets will be dealt with in the next chapter.

APPENDIX: AN EXAMPLE OF NUMERICAL ANALYSIS

To illustrate the solution of gravitational problems by numerical analysis we consider a simple problem. A body is projected vertically with a speed of 1000 ms^{-1} from the surface of the moon. What will be its greatest altitude? Newton's law of gravitation tells us that the acceleration of the body is given by the formula $f = - GM/r^2$, where G is the gravitational constant, M is the lunar mass and r is the distance to the centre of the moon. The value of f will decrease with altitude since r = (altitude + lunar radius).

The essence of the numerical method is the use of small time steps. Instead of trying to compute directly the maximum altitude, we compute a sequence of altitudes at different times. If the step in time is small enough, we can assume that during a particular step the change in acceleration is negligible since the body moves only a relatively small distance.

Let us choose a time step of 50 s. We know that at time $t = 0$ s the speed $v = 1000$ ms^{-1}, and the acceleration

$$f = \frac{-4.9 \times 10^{12}}{(1.738 \times 10^6)^2} = -1.62 \text{ ms}^{-2} .$$

We attempt now to find the position and speed at $t = 50$ s. Since we assume constant f during the time step, the speed will be reduced by $50 \times 1.62 = 81$ ms^{-1}, i.e. $v = 1000 - 81 = 919$ ms^{-1}. In addition, we take the distance travelled to be the time multiplied by the average speed during the step, giving $50 \times \frac{1}{2}(1000 + 919) = 47975$ m. This completes the first time step; after 50 s the speed and height are 919 ms^{-1} and 47975 m respectively.

The speed and height at $t = 100$ s are estimated by computing a second step of 50 s. The acceleration must be recalculated because it will be reduced. Using the same formula,

$$f = \frac{4.9 \times 10^{12}}{(1.738 \times 10^6 + 47975)^2} = -1.54 \text{ ms}^{-2} .$$

The lower deceleration results in a speed reduction of 77 ms^{-1} and the lower speed gives a smaller increase in height of 44025 m. Thus after 100 sec we obtain $v = 842$ ms^{-1} and height 92000 m. The computation is continued until the value of v becomes negative, indicating that the body is falling back towards the surface of the moon. With the time step of 50 s the maximum height indicated is 366000 m.

The result obtained above is not precisely correct. This will be clear since the assumption of constant acceleration within a step is not strictly true. However, the reduction of step-size will make the approximation a better one. Table 6.2 shows the solution obtained with four different time steps.

Table 6.2

Step (s)	Maximum height (m)
50	366 000
25	370 000
10	373 000
2	374 350

Of course a reduction in step-size means that the computation takes longer but the extra accuracy may justify this. In this case we can obtain the exact solution from energy considerations and to the nearest metre the maximum height would be

376 477 m. The table shows that a reduction in step-size by factor of 25 causes a reduction in error by a similar factor.

It should be stressed that the numerical method illustrated here is a very crude one which was chosen for its simplicity. Methods used in practice, with the aid of computers, are capable of yielding high accuracy without requiring very small step-sizes.

7

Planetary condensation

7.1 INTRODUCTION

The problem of planetary condensation is perhaps the most difficult one in solar-system cosmogony. It is necessary to describe a mechanism by which the planets may be assembled from diffuse material. In the Capture Theory the material is derived from a protostar tidally disrupted by the sun, and so it is not obvious that the solar gravitational field will allow it to recondense. Even so the theory presents less difficulty in this respect than the nebula theories since, following capture, proto-planetary material should be of similar density to that of the protostar. A similar mass distributed evenly with solar-system dimensions would be much more rarefied. Some of the ideas proposed by the nebulists have been discussed in Chapter 3 but it is useful here to summarize the factors relating to condensation.

The formation of major planets as a result of random accumulation of small bodies seems to be ruled out by the excessive timescale for such a process, although terrestrial planets present a lesser problem. The problem seems similar to that of assembling a planet from the asteroid belt, or forming a new satellite from the Saturnian rings. Actually, at the present time, dispersive forces would prevent either of these processes from occurring; Jupiter has a perturbing gravitational effect on the asteroids, and, of course, the rings of Saturn lie within the Roche limit for tidal disruption. Such forces would not be operative within a primitive nebula and so some cosmogonists have suggested that the asteroid region today exhibits the types of bodies which would have inhabited the primitive solar system prior to the accumu-lation of the major planets. This idea has motivated exotic theories for the direct formation of small bodies as the building blocks of planets. In fact it is much easier to comprehend the formation of large gaseous bodies, in a sufficiently dense environ-ment, from which smaller, rocky ones might then arise in the appropriate circum-stances. As we shall see later in Chapter 13, there is some evidence to suggest that meteorites, and presumably asteroids, once formed parts of larger bodies.

In this chapter we shall examine planetary condensation in stages. First we consider the evolution of an isolated, mainly gaseous body (called a protoplanet). Next we look at how several such protoplanets could develop from the material captured by the sun. Finally, the effect of external tidal forces, mainly due to the sun, will be assessed by modelling a protoplanet in orbit.

7.2 ISOLATED BODIES

The Capture Theory predicts that protoplanetary material would have a density of approximately 10^{-8} kg m^{-3} and a temperature in the range 10–100 K. The composition would be that of a typical star, i.e. mainly hydrogen. Since the material has been drawn from a more massive body in a state of collapse, it is clear that a sufficient quantity of it is capable of withstanding dispersive forces. In other words the gravity of the protostar dominates the internal pressure force thus causing it to contract. A body in this state is said to be gravitationally unstable. In 1919 Jeans considered in detail the problem of determination of the critical mass for instability, called the 'Jeans' mass in recognition of the work. For a sphere of molecular hydrogen with density ρ and temperature θ the Jeans mass (see equation (1.2)) can be expressed as

$$M_J = 2.7 \times 10^{21} \sqrt{(\theta^3/\rho)} \qquad (7.1)$$

We can see from this formula that the mass increases with increasing temperature and decreases as the density increases. This is intuitively reasonable since the internal pressure will increase as temperature increases, making the gaseous body tend to expand. The formula (7.1) can be expressed in a different way. If a mass is specified then the density of material necessary for gravitational instability is given in terms of temperature, i.e.

$$\rho = 7.29 \times 10^{42} \ \theta^3/M_J^2 \qquad (7.2)$$

In Fig. 7.1 there is shown the variation in this critical density with temperature for some particular planetary masses. It is clear from this that the direct formation of bodies of mass comparable to the terrestrial value is not likely in the Capture Theory. With a temperature $T = 30$ K and density 4×10^{-8} kg m^{-3} the critical mass is about one Jupiter mass.

Consideration of the density distribution of captured material (Fig. 6.20) may give the impression that formation of any planets other than Jupiter by means of gravitational instability is unlikely. The situation is not quite as bad as this, for the Jeans mass is that value dividing bodies which would *start* collapsing or *start* disrupting. For a body with less than the critical mass, surface material would start to move outwards, while, in the absence of pressure gradients in a homogeneous body, interior material would start to move inwards under gravity (see Fig. 7.2). Since the boundary disturbance can travel inwards only at the velocity of sound there may be some interior surface which, as a result of the initial collapse, will contain more than the relevant Jeans mass before the disturbance reaches it. Such a surface would form the boundary of an unstable core containing some fraction of the mass of the original body.

A protoplanet with mass greater than the critical value would collapse significantly in a time comparable to the 'free-fall' value as given by equation (5.5). From this we find that the free-fall time for our example above is about 10 years. Note that this assumes no forces act other than gravity; an actual collapse would take longer and be halted eventually by the build-up of pressure gradients and by energy losses.

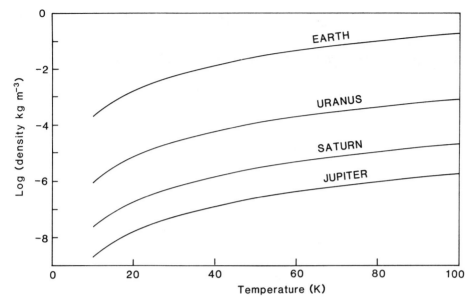

Fig. 7.1 — Variation of critical density with temperature for various planetary masses (as Jeans mass).

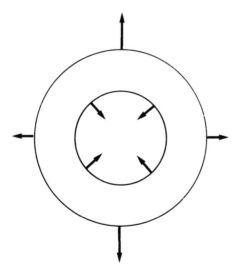

Fig. 7.2 — Initial behaviour of a protoplanet. Outer layers expand while inner ones contract.

It is important here to realize that the collapse times for Capture Theory proto-planets may be much shorter than their initial orbital periods. Consequently it is possible that a protoplanet may have collapsed almost completely by the time it makes its first perihelion passage following capture (see section 6.6).

7.3 EARLY EVOLUTION OF A PROTOPLANET

In 1982 Schofield and Woolfson published the results of a detailed study of protoplanetary evolution. At first a mathematical model of an isolated protoplanet was constructed. Of course the protoplanet would be subject to gravitational effects due to the sun and the receding protostar. These would certainly destroy the symmetry and thus render the modelling problem more difficult. An isolated body would remain spherically symmetric and permit physical effects such as heat flow to be modelled conveniently and realistically. The model occupies only one space dimension instead of three and so the reduction in computational complexity is considerable. The analysis of such a model provides useful information which can be applied later to the asymmetric body.

For numerical purposes the model is discretized, i.e. it is split into a moderate number of separate parts. In this case the parts are concentric spherical shells (Fig. 7.3) of equal mass and each characterized by pressure, density, temperature and

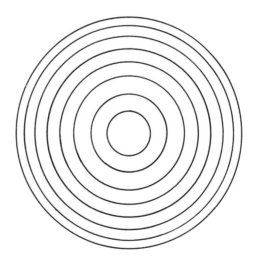

Fig. 7.3 — Model construction in the form of spherical concentric shells.

chemical composition. Initially the protoplanet has a very low density (by terrestrial standards) and so heat flow is considered to be due only to radiation, which is dependent on the opacity of the gaseous material. Like our present-day atmosphere, the planetary material would be fairly transparent. Transport of heat by conduction and convection in such a medium would not be significant, and so these factors are neglected.

The character of the model restricts its validity to the very early stages of planetary evolution; today's major planets are definitely opaque and display considerable evidence of convective processes. The restriction is not a serious drawback because the objective here is to discover whether or not captured material is capable of collapsing into dense planetary bodies. The problem of evolution or equilibrium at high densities is a different one, and one that is not of prime importance from the cosmogonist's point of view.

The composition of the protoplanet is taken to be 74% hydrogen, 24% helium and 2% of heavier elements and to be initially homogeneous. Also the material behaves like an ideal gas satisfying $P = \rho k\theta/\mu$, where k is Boltzmann's constant and μ is the mean molecular weight. The model parameters are shown in Table 7.1.

Table 7.1 — Protoplanet model

Initial mass	2×10^{27} kg $\equiv 1$ Jupiter mass
Initial radius	2.285×10^{11} m ≈ 1.5 AU
Initial density	4×10^{-8} kg m^{-3}
Composition:	Hydrogen 74%
	Helium 24%
	Other elements 2%
Mean molecular weight	3.88×10^{-27} kg
No. of shells	16
Initial temperature	
(a) Jeans mass (critical)	32.8 K
(b) 1.5 Jeans mass	24.6 K

The model was used to simulate the evolution of a protoplanet in a number of cases. The ratio of the mass to the critical mass was varied by changing the initial temperature; thus $\theta = 32.8$ K makes the Jupiter mass precisely critical (equation (7.1)) while $\theta = 24.6$ K corresponds to the same mass being 1.5 times the critical value. With the lower temperature, gravitational forces would dominate the pressure forces and so we would expect rapid collapse. In addition some further experiments were made with a protoplanet embedded within an external medium exerting a small pressure at the surface of the body.

Some of the results of the simulations are illustrated in Figs 7.4–7.6. The first of these shows the evolution in the case where the initial temperature is critical. As in Fig. 7.2, the outer regions expand as the interior collapses. After about nine years the disturbance reaches the centre causing a 'bounce' and subsequent re-expansion lasting approximately 70 years. Following this there is a period of slow contraction, lasting around 500 years, during which the central density and temperature reach 3.4×10^{-6} kg m^{-3} and 180 K respectively. Then the evaporation of water ice from grains reduces the opacity and absorbs latent heat, thus promoting a further sustained collapse during which higher temperatures lead to H_2 dissociation at about

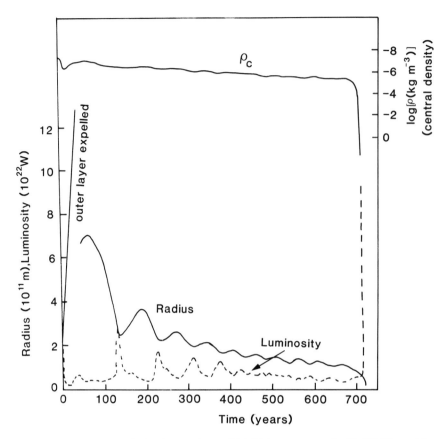

Fig. 7.4 — Evolution of an isolated protoplanet with initial temperature $\theta = 32.8$ K (Planet (a) from Table 7.1). Mass is 1 Jupiter mass. (Schofield and Woolfson, 1982.)

1500 K. This causes an even more rapid collapse, attaining near free-fall character in the interior. After a total time of about 720 years the central density and temperature reach 24 kg m^{-3} and 6000 K respectively. These are limiting values for the model, and so the simulation must be terminated. It should be noted that the outer layer of the model attains escape velocity and is lost completely.

In contrast, the cooler protoplanet collapses more quickly, attaining similar central density and temperature after only 20 years. Although the outer layers of this model also expand, they do not achieve escape velocity. Fig. 7.5 illustrates the physical state of this protoplanet after 20 years.

Intermediate starting temperatures result in intermediate collapse times; the variation is plotted in Fig. 7.6. As might be expected, the effect of an external medium exerting a pressure on the protoplanet is to make the collapse more rapid for a given initial temperature. Also the ejection of outer layers is prevented.

Although the above simulations have neglected completely any external gravitational and heating influences, they do yield important limits on planetary condensa-

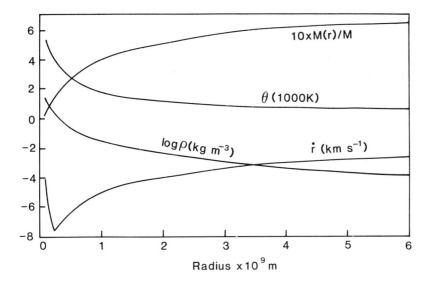

Fig. 7.5 — State of protoplanet (b) (Table 7.1) at end of simulation. Only the inner region comprising about 65% of the total mass is shown. Expansion of surface has resulted in a planet radius of more than 4 AU but the outer atmosphere is very tenuous.

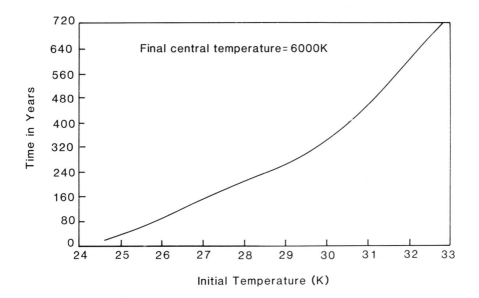

Fig. 7.6 — Variation of collapse time of a protoplanet with initial temperature.

tion in the Capture Theory. We recall from section 6.6 that protoplanetary material is captured near perihelion into very eccentric elliptic orbits. Thus, following removal from the protostar, it moves rapidly away from the sun and outside its Roche limit, and under appropriate conditions of temperature and density, there will be significant condensation before the next perihelion passage. Since the orbital period is of order 100 years, it is seen from the above experiments that a protoplanet must possess about 1.5 times the Jeans mass otherwise it will still be too tenuous at the second perihelion to withstand solar tidal stress. If the protoplanet does collapse rapidly then it may well reach present-day densities during its first orbit — if not it will surely suffer further disruption and it is not certain that a condensed body will form.

It is not possible effectively to model heating by solar radiation when spherical symmetry is imposed. Nevertheless Schofield and Woolfson (1982a) computed the evolution by assuming the whole of the planet was bathed in solar radiation (assuming present-day solar luminosity). This increases the probability of dispersion, but it turns out that the cooler protoplanet would still condense in a very short time compared with the orbital period. Realistic modelling of tidal effects is not possible with the symmetry restriction and so we now consider a three-dimensional treatment.

7.4   CONDENSATION IN A TIDAL FIELD

To examine the evolution of a protoplanet in the gravity fields of the sun and the departing protostar special mass-point models have been developed. These are similar to the one used for protostar simulation (section 6.6). The material of the protoplanet is represented by a large number of point masses filling its volume. Equations of motion for each mass point are solved numerically using a computer. This is a step-by-step process (see Appendix) and thus provides a snapshot of the state of the protoplanet at each time step, rather in the manner of a cine film. To start the simulation the spherical protoplanet is placed on a heliocentric elliptic orbit whose elements are suggested by the encounter studies. Fig. 7.7 shows some typical results computed by Dormand and Woolfson in 1971.

A defect of such models is that they cannot be made to simulate the physical properties of a protoplanet as realistically as the spherical shell model described above. However, the effect of pressure forces and viscosity can be approximated crudely by the inclusion of repulsion forces to oppose point-mass gravity, and also by smoothing the point velocities. Thus it is possible to construct a model which will, in isolation, collapse significantly in about twice the free-fall time of the shell model (section 7.3). With this property determined, we can have some confidence in the validity of the model when external forces are applied. As we have indicated earlier, the most important factor determining the fate of a protoplanet (given an initial orbit) is the starting density. If this is too small then the protoplanet is ripped apart by the tidal forces due to the sun. A high-density start will lead to condensation, and there are intermediate states involving degrees of partial disruption.

Schofield and Woolfson (1982b) were able to match fairly well the behaviour of their mass-point model with that of the physically realistic spherically symmetric case (section 7.3). They also introduced a novel scheme to alleviate one of the most disabling features of mass-point dynamics — the pooling problem. If two or more

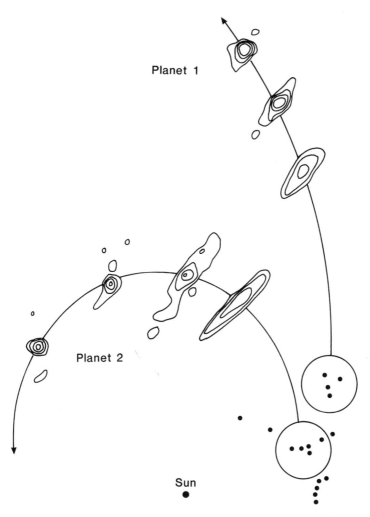

Fig. 7.7 — The evolution of two protoplanets suggested by a limacoid simulation (see Fig. 6.14). The relation is shown between the filament mass-points and the initial planetary mass-point configurations (circles). The contours indicate densities in the plane of the encounter orbit. The outermost contour marks a minimum density of $1.7 \times 10^{-7}\,\mathrm{kg\,m^{-3}}$. Inner contours indicate two, four, and eight times this value (Dormand and Woolfson, 1971).

points make a close approach, the accuracy of the computational procedure will be reduced unless the time step is reduced. Since the cost of the calculation is the same for each step, whatever its length, this means that the overall computing time increases, sometimes to an unacceptably large extent. If a protoplanet is collapsing the step length must be reducing and so it may be impossible to continue the calculation for a significant part of the orbit due to the exponentially falling step size. In the encounter simulation the 'limacoid' technique (section 6.6) was employed to reduce the length of the computation. The above authors developed a mass redistribution scheme which could be used to regularize the point spacings at suitable stages (Fig. 7.8). Conservation of energy, momentum, and angular momentum were

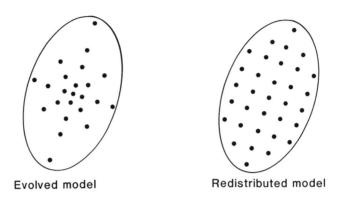

Evolved model   Redistributed model

Fig. 7.8 — An illustration of the concept of redistribution (Schofield and Woolfson, 1982.)

ensured by varying the masses of the regularized points and computing suitable new velocities.

The new technique yielded a useful extension in the point-mass scheme. It was applied to a protoplanet with the properties (in isolation) of the spherical model (section 7.3) and confirmed the likely condensation of such a body. An important parameter derived from such simulations (Fig. 7.9) was the planetary rotational angular momentum. As we have seen in section 6.4, a body moving in a non-uniform gravitational field will suffer a torque which yields an angular acceleration. Viewed another way, the tidal tip of a protoplanet tries to point to the tide-producing body (sun) (Fig. 6.9), and hence the orbital motion causes the body to rotate. The simulations showed that it was possible for the Capture Theory to predict a planetary angular momentum distribution reminiscent of the Jovian system. This has important implications for satellite formation, which will be the subject of the next chapter.

7.5   SPH SIMULATIONS

Although the Schofield–Woolfson planetary modelling was a considerable advance on previous work, we can identify a few defects. The point regularization scheme was applied within an ellipsoid, thus making it impossible for the protoplanet to take on any other shape. Also the non-gravitational forces were physically based only in that they caused the protoplanet to condense properly in isolation. These problems do not invalidate the results obtained but, generally speaking, we would be more confident if our models contained only 'real' physical constants. This is extremely difficult to achieve, but we described in section 6.7 the SPH technique which, although possessing many of the convenient aspects of mass-point schemes, permits realistic modelling of pressure forces according to an equation of state. The method eliminates the potential singularities of mass points, even though condensation will

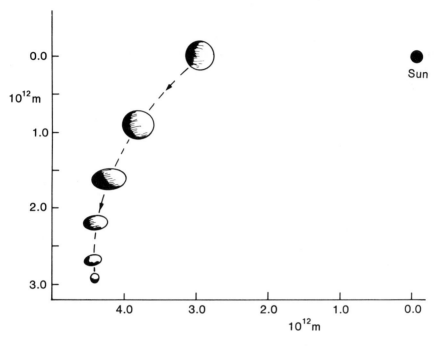

Fig. 7.9 — Protoplanetary evolution starting from the semi-minor axis ($a = 20$ AU, $e = 0.77$) (Schofield and Woolfson, 1982).

still reduce the length of time steps. The computational cost of the SPH technique is high, as is the cost of mass-point simulations, because large numbers of elements are desirable to achieve acceptable resolution. The requirement for each element to affect each other element implies that the cost depends on the square of the total number of elements. However, there will be no need to define any shape for the purpose of regularization (sections 6.6, 7.4).

Before describing some of the results from SPH simulations it is important to recall more recent results from Capture Theory encounters. Figs 6.18, 6.19, and 6.20 refer to the capture of material from a collapsing protostar. The total mass captured was about $M_\odot/25$, or about 28 times the total mass of the present-day planets, and the densities were somewhat higher than had been predicted earlier (Fig. 6.20). The sequence in Fig. 6.18 indicates the break-up of a tidal filament similar to the prediction of Jeans; each of the fragments is much more massive than any known planet. Thus it seems reasonable to investigate planetary condensation starting with initial conditions different from those of Table 7.1.

Coates (1980) and Allison (1986) have developed SPH models suitable for our problem. In 1988 Dormand and Woolfson reported the simulation of an initially homogeneous protoplanet of three Jupiter masses and initial density 5×10^{-8} kg m^{-3} with a temperature 15 K. This was placed on an orbit the same as that in Fig. 7.9. The period was about 90 years and the perihelion distance was about the same as that of Jupiter.

The high mass assumed here is not inconsistent with solar-system parameters. The total mass captured as a result of the encounter described in section 6.7 was about 40 times that of Jupiter, but, as we shall see in Chapter 9, the present system may be considered to be the residue of a much larger quantity of material. If much mass is to be lost then it is certainly necessary to analyse the behaviour of a 'protoplanet' more massive than any known to exist at present. It will be seen (Fig. 6.20) that the initial density corresponds well with higher densities observed in material captured from the rapidly collapsing protostar (Fig. 6.18).

Although the filamentary material being modelled here is termed a 'protoplanet', there is no strong evidence to suggest that, at the stage being considered, the differentiation of the filament to form planetary condensations has begun. The resolution of our SPH model is not yet high enough to provide this information. An approach intermediate between the one here and that which models the complete protostar might involve the whole of the captured material (with SPH), with the sun and the receding star as point masses. This looks attractive, but there would be difficulties in specifying initial velocities for the whole of a filament, and, of course, the resolution would still be low. If around 100 fluid elements were employed, then each would represent a mass greater than that of Saturn.

The approach yielding optimal resolution might involve a planet of Jupiter mass. Yet this could provide misleading results when there is, as a result of the earlier experiments, much more material available to influence planetary condensation. This factor could be extremely significant since the critical density for Jeans mass decreases as the mass increases (Fig. 7.1). It seems quite possible that there will be circumstances in which a protoplanet initially of one Jupiter mass will be tidally disrupted, but a more massive body will give rise to a Jovian planet as a consequence of partial disruption.

Six snapshots of the SPH protoplanet, at times up to 19 years from the start, are shown in Fig. 7.10. The points plotted are the projected positions of the SPH fluid elements with respect to the central element. Initially the protoplanet was within the Roche limit but moving rapidly away from the sun. Up to 10 years the shape is fairly ellipsoidal, but afterwards the shape may be better described as that of a cigar or a filament. From the density of the elements it is seen that condensation is taking place and after 19 years the highest densities are to be found near the extremities of the object. It appears that at least two, and possibly four condensations (planets?) are growing. These are not gravitationally bound and further tests show that the extreme centres of condensation will continue to diverge. So we see that the protoplanet has been disrupted *but not dispersed completely*. Again the pictures are indicative of the Jeans fragmentation theory of filaments (section 1.3). Fig. 7.11 shows the contours of density for the protoplanet(s) after 19 years. At this time the planet has not reached aphelion and so it is still moving away from the sun.

7.6 CONCLUDING REMARKS

The experiments described in this chapter have confirmed the plausibility of the capture hypothesis. It must be said that a set of parameters appropriate to the actual solar system has not been found. However, the idea of disrupted material recondens-

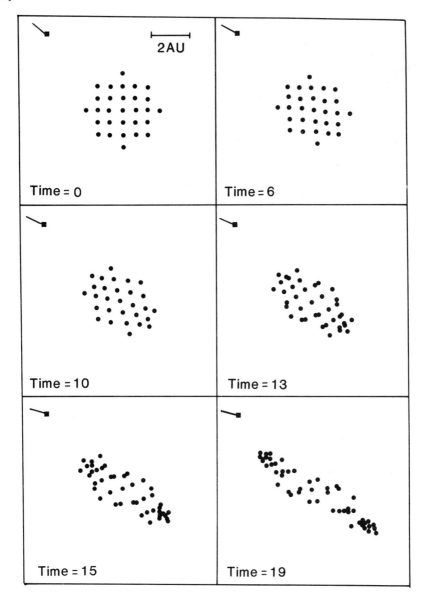

Fig. 7.10 — A sequence of configurations of a model protoplanet showing disruption and condensation. Dots indicate projected positions of fluid elements. Vector at top left shows position of planet with respect to the sun. times are in years. (Dormand and Woolfson, 1988).

ing after being captured seems reasonable, although the mechanisms here apply to the formation of major planets and not directly to terrestrial-type bodies.

The evolution of a protoplanet has been considered for only a very short time in comparison with the known age of the solar system, but it is safe to assume the survival of any body which displays a significant degree of condensation. When a

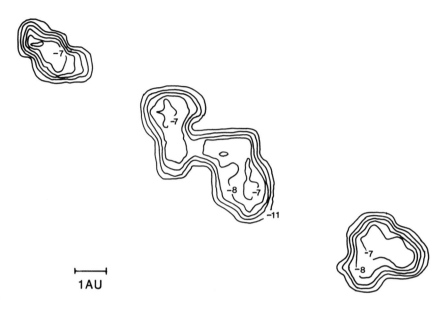

Fig. 7.11 — Density contours in a disrupted protoplanet after 19 years. Labels indicate log (density/kg m^{-3}). (Dormand and Woolfson, 1988.)

rapid collapse has set in it will be completed very quickly indeed — perhaps in days rather than years. Of course the kinetic energy of contraction has to be dissipated before the planet is left in a stable condition and so we expect the new planet to be very luminous indeed for a short time.

The current state of modelling the capture event and associated planetary formation is far in advance of that available when the hypothesis was formed, but improvements are still needed to give a better insight into the physical processes involved. More continuity is desirable. The gap between protostar disruption and protoplanet evolution has not yet been properly bridged. In part such an improvement depends on available computing power, since resolution is still fairly low, but more advanced mathematical techniques are expected to play an important role in the future.

The problem of planetary satellite formation obviously depends on planetary condensation, but, such is the gap between the masses of the two types of object, no such formation can be inferred directly from the above calculations. We explain satellite cosmogony, related to present ideas of planetary formation, in the next chapter.

8

Satellite formation

8.1 REGULAR AND IRREGULAR SATELLITES

When Galileo first observed the four major satellites of Jupiter in 1609 he was greatly reinforced in his belief in the Copernican system. It was already known that the moon was an earth satellite, but here, in small scale, was a system for all to see which exactly modelled the Copernican concept. Since that time many other satellites have been observed and the characteristics of their orbits measured, and there is no reason to believe that other discoveries will not be made from time to time.

In Table 2.3 there is shown the current state of knowledge concerning planetary satellites. From this it will be seen that the innermost major planets, Jupiter, Saturn and Uranus, are those with the major satellite families. In addition, the satellites seem to fall into two classes, the first being those closest to the planet with near-circular prograde orbits in the planetary equatorial plane and the second being those which are less regular in their orbital characteristics. These will be referred to as the *regular* and *irregular* satellites respectively.

It is possible that satellite families could form as irregular systems, perhaps as individually captured bodies, and that the regularity of the close-in satellites is due to evolutionary factors dependent on tidal effects, for example. Convincing mechanisms for the capture of individual satellites and subsequent evolution to a regular system have not been advanced, but it could not be said with complete certainty that such a process is impossible. However, the generally accepted view is that the characteristics of regular systems have something to do with their origin and this is the assumption we shall accept here.

Galileo's observation, that the major-satellite system of Jupiter resembles the planetary system in miniature, is often taken to indicate that the two types of system must have originated through similar mechanisms. Indeed to some people this idea has become an article of faith and the view of Jeans in this respect has already been given in section 4.1. More recently Alfvén (1978) has stated that 'We should not try to make a theory of the origin of planets around the sun but a *general theory of the formation of secondary bodies around a central body*. This theory should be applicable both to the formation of satellites and the formation of planets.'

Although the first-glance similarity of the planetary system and the regular satellite systems is fairly striking, it is worthwhile examining the characteristics of these systems in somewhat greater detail. In particular we should recall the angular momentum problem which bedevilled the earliest ideas about the origin of the solar system. One way of expressing this problem is to look at the ratio of the intrinsic angular momentum (angular momentum per unit mass) for planetary material due to its orbital motion to that of solar material due to its spin. This ratio will be of order

$$S = \frac{(GM_{\odot}r_{\mathrm{P}})^{1/2}}{R_{\odot}^2 \omega_{\odot}} \tag{8.1}$$

where $M_{\odot}$, $R_{\odot}$ and $\omega_{\odot}$ are the mass, radius and spin angular velocity of the sun, and r_{P} the radius of a planetary orbit. Taking r_{P} as the mean radius of Jupiter's orbit (the extreme planets would give a factor ~ 3 greater or less for S) we find $S \approx 7800$. The angular momentum problem is then seen to be that of separating from a common pool of material some small part with intrinsic angular momentum nearly four orders of magnitude greater than that for the remainder — indeed to the extent of extracting virtually all the angular momentum in a tiny fraction of the mass.

If the value of S is now calculated with Jupiter and Io as the primary and secondary bodies respectively, then we find a very different situation, with $S = 8$. In Table 8.1 various values of S are shown for various pairs of bodies in the planetary and satellite systems. It is clear that, whatever its other faults, the Laplace nebula theory would have had no angular momentum problem in explaining the regular satellite systems.

Table 8.1 — Ratio of intrinsic angular momentum of secondary orbit to that of primary spin at equator

Primary	Secondary	Ratio
Sun	Jupiter	7800
Sun	Neptune	18 700
Jupiter	Io	8
Jupiter	Callisto	17
Saturn	Titan	11
Uranus	Oberon	21

The remainder of this chapter will be devoted to an alternative model for the formation of satellite systems which follows naturally from the Capture Theory proposals for the formation of the planets themselves.

8.2 TIDAL EFFECTS ON EARLY PROTOPLANETS

In Chapter 7 a model of an evolving protoplanet was described which took into account, in great detail, the physical properties of the protoplanetary material. If a planetary mass of material, taken from the collapsing protostar, was cool enough then it would form a comparatively dense core and collapse overall on a timescale of some tens of years. The Capture Theory model shows that the initial orbits of planets would have been highly eccentric and the orbit of a proto-Jupiter, for example, would have had a period approaching 100 years. With planetary material being released from the protostar after the perihelion and moving towards the aphelion of the initial planetary orbit, it seems probable that the density of at least the central region of the protoplanet will be high enough to resist disruption at the first perihelion passage. For example, Schofield and Woolfson (1982a) showed that protoplanet (b) in Table 7.1 would have collapsed on a timescale of 20 years to give a central temperature and density of 6000 K and 33 kg m^{-3} respectively and a total radius of 4 AU, but to be so centrally condensed that one half of the total mass is contained within 0.003 of the total radius (0.013 AU). The distribution of density and contained mass is illustrated in Fig. 7.5.

   The indication of these results is that at and near the first perihelion passage the protoplanet will be an extended body with a high central condensation. Its ability to resist disruption will be governed by the Roche limit (equation (3.3)) which will here be used in the more precise form

$$R_1 = \left\{ \frac{6M_\odot}{\pi\rho_m} \right\}^{1/3} \tag{8.2}$$

where R_1 is the minimum distance from the sun for a body of mean density ρ_m to remain whole. This may also be written as

$$\rho_m = \frac{6M_\odot}{\pi R^3} \tag{8.3}$$

where ρ_m is the minimum mean density required to preserve the integrity of a protoplanet at a distance R from the sun.

   For one model, suggested by the Capture Theory and described in more detail in Chapter 9, the initial proto-Jupiter has an orbit with semi-major axis 14.1 AU and eccentricity 0.80 (Table 10.1) leading to a perihelion distance of 2.9 AU. At this distance the critical mean density, ρ_m, from equation (8.3), is 4.55×10^{-5} kg m^{-3}. If all parts of the protoplanet have a lesser density than this then, theoretically, it would be disrupted by the perihelion passage. However, if the density is not too far below the critical density then it might totally or partially resist disruption, depending on how quickly it passes through the perihelion region. This is what happens, for example, in the Capture Theory calculations where the protostar passes within the Roche limit but retains its identity, if not integrity, after the tidal interaction. An important conclusion which we can reach here is that if ρ_m from equation (8.3) falls somewhere between the central density of the protoplanet, ρ_c, and the average density of the protoplanet then there will be some spherical surface for the undistorted protoplanet within which the mean density is ρ_m. The material inside this surface should be retained; that outside may be lost, but probably only in part because of the transient and variable nature of the tidal forces.

The Capture Theory model gives the pattern that, as successive planets are formed, there are monotonically decreasing semi-major axes and increasing eccentricities. This combination, corresponding to ever-decreasing periods and perihelion distances, is progressively less conducive to planetary formation. There will be some minimum distance below which no planet will be able to form and some early work suggested that this limit was somewhere in the region of Mars (Dormand and Woolfson, 1971).

In Table 8.2 there are shown the postulated initial orbits for Jupiter, Saturn and Uranus. The values of ρ_m are calculated from equation (8.3) with R equal to the perihelion distance. With the assumption that the retained mass is a little greater (around 5%) than the present planetary mass, it is possible to calculate the radius, R_c, of the limiting spherical surface which contains that part of the material which will eventually form the planet . A model for the dynamical behaviour of this retained mass, regarded for our modelling purposes as a distorted originally spherical body, will now be considered.

Table 8.2 — The initial orbits of protoplanets and the deduced average density and radius of the spherical surface of the retained material

Planet	Initial mass (kg)	Initial semi-major axis (10^9 km)	Initial eccentricity	ρ_m (kg m^{-3})	R_c (10^7 km)
Jupiter	2.00×10^{27}	2.19	0.80	4.55×10^{-5}	2.19
Saturn	5.97×10^{26}	2.79	0.68	5.37×10^{-6}	2.98
Uranus	9.25×10^{25}	5.34	0.69	8.42×10^{-7}	2.97

The general behaviour of the protoplanet as it approaches perihelion may be inferred from the tidal-interaction calculations which have been done for the Capture Theory model. As the protoplanet approaches the sun so it would develop a tidal bulge which would lag behind the radius vector to the sun. This lag is α in Fig. 8.1. Under most conditions the time for collapse of the protoplanet will be considerably greater than the time for a perihelion passage (between one and two years to go from one semi-latus rectum to the other) and so the model is assumed to have a fixed tidal distortion for this period. Since the planet has a high central condensation, the form of the critical equipotential surface, within which the material will resist disruption, can readily be found. For a mass ratio of sun : protoplanet of 1000 : 1, the cross-section of the equipotential surface is shown in Fig. 8.2; for high mass ratios the distance from the tidal tip to the centre of mass is about 1.4 times the original radius of the undistorted body.

It is clear that the differential forces due to S at the points O and P in Fig. 8.1 will

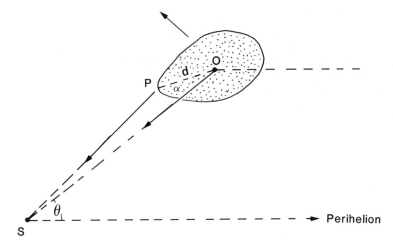

Fig. 8.1 — Forces acting on an extended protoplanet with a tidal lag.

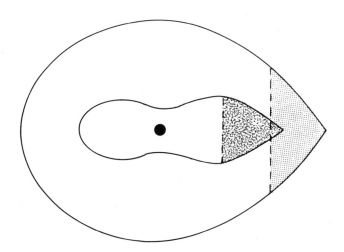

Fig. 8.2 — The outer profile shows the cross-section of a tidally distorted protoplanet for a sun : protoplanet mass ratio of 1000 : 1. The inner profile shows the cross section after free-fall collapse for a time.

impart spin angular momentum to the protoplanet. A system of equations can be set up describing the way in which the spin angular velocity of material in the tidal bulge varies with time, and these calculations have been done for a range of models by Williams and Woolfson (1983). The results obtained depend on the assumed initial conditions, including initial spin of the protoplanet. In the model for the collapse of a proto-Jupiter described by Schofield and Woolfson (1982b), a considerable amount of spin angular momentum was imparted to the protoplanet well before it reached the perihelion region of its orbit. A very pleasing aspect of the calculations by Williams and Woolfson is that a wide range of initial conditions, encompassing all physically plausible situations, gives remarkably small variations in the final result. The final spin angular velocities, ω_f, of tidal-bulge material for the planets with natural satellites are calculated as

Jupiter	$\omega_f = 8.02 \times 10^{-8} \, s^{-1}$
Saturn	$\omega_f = 2.66 \times 10^{-8} \, s^{-1}$
Uranus	$\omega_f = 1.06 \times 10^{-8} \, s^{-1}$

with a possible variation of $\pm 20\%$.

We must now deduce what these results mean in terms of planetary spin rates and satellite formation.

8.3   THE ANGULAR MOMENTUM OF THE TIDAL BULGE

The spin angular velocities given at the end of the last section relate to the tidal tip of the distorted protoplanet. However, it is clear from Fig. 8.1 that at any instant the sense of the torque on any element of matter, i.e. whether it is prograde or retrograde, will depend on which side it is of the line OS. Since the protoplanet will be very centrally condensed, with most of the mass concentrated in the region around O, it will readily be appreciated that the majority of the imparted angular momentum will be concentrated in the tidal bulges and most of that in the larger sun-facing bulge. The total final angular momentum contained in the major bulge depends on a number of factors:

(1) the volume of the bulge material, V_b;
(2) the ratio of the mean bulge density to the mean density for the whole protoplanet;
(3) the distribution of density within the bulge;
(4) the distribution of angular speed within the bulge material.

A numerical study of the profile shown in Fig. 8.2 reveals that the major bulge contains about 10% of the total volume of the protoplanet so that, approximately

$$V_b = 0.40R_c^3 \tag{8.4}$$

The other factors are difficult to assess without detailed modelling, which has not yet been done, but we may certainly write for the total angular momentum of bulge material

$$J_b = 0.40f\rho_m R_c^5 \omega_f \tag{8.5}$$

where f, which allows for all the uncertain factors, (2), (3) and (4) above, is certainly very much less than unity.

It will be seen in Fig. 8.2 that the smaller tidal bulge is by no means negligible, and for the orbits given in Table 8.2 it is found numerically that the ratios in the volume of the two bulges are 0.35, 0.50 and 0.65 for Jupiter, Saturn and Uranus respectively. Taking into account the lesser extension of the smaller bulge, we estimate for Jupiter that the total angular momentum in the two tidal bulges may be given as

$$J_{2b} = 0.50 f \rho_m R_c^5 \omega_f \qquad (8.6)$$

where the f may be slightly different from that in equation (8.5). The numerical factors for the other two planets, in place of 0.50, are 0.55 and 0.60.

Inserting in equation (8.6) the values found for ω_f together with ρ_m and R_c from Table 8.2, there are obtained for the three planets the values of J_{2b}/f given in the first column of Table 8.3.

Table 8.3 — Comparison of theoretical induced angular momentum in tidal bulges with spin angular momenta of planets and orbital angular momenta of satellite systems. The augmented values assume that the present satellites are 2% solid residues of original gaseous protosatellites

Planet		Observed	Orbital AM	Augmented AM	Expected total
	J_{2b}/f	spin AM	satellites	satellites	initial AM
Jupiter	9.2×10^{39}	4.4×10^{38}	4.2×10^{36}	2.1×10^{38}	6.5×10^{38}
Saturn	1.8×10^{39}	9.1×10^{37}	9.5×10^{35}	4.7×10^{37}	1.4×10^{38}
Uranus	1.2×10^{38}	1.7×10^{36}	1.3×10^{34}	6.5×10^{35}	2.4×10^{36}

All units are kg m^2 s^{-1}.
AM, angular momentum.

8.4   THE MECHANISM OF SATELLITE FORMATION — A GENERAL DESCRIPTION

After departing from the perihelion region, the protoplanet, distorted and possessed of spin angular momentum, will continue its collapse evolution. This will involve the complex interplay of forces and factors as described by Schofield and Woolfson (1982b), with the gravitational forces inducing collapse and meeting increasing resistance from gas-pressure forces as the opacity of the material and internal temperatures increase. We shall now consider some extreme conditions which will certainly not prevail for the protoplanet we are considering.

If the protoplanet had reached some kind of near-equilibrium configuration, so that it was neither expanding nor contracting very quickly, then it would eventually evolve to a spherical shape after oscillating and dissipating mechanical energy. This assumes a near-balance in pressure and gravitational forces. Another extreme case would be where the gravitational forces were so dominant that the distorted planet was in a state of free-fall, or near-free-fall, collapse. The general pattern of such a collapse is that it is a runaway effect with very little apparently happening for an appreciable fraction of the free-fall time. It also gives the characteristic that any departures from spherical form of the initial configuration are greatly exaggerated by the collapse. This is illustrated in Fig. 8.2 where the profile of the protoplanet is shown after a considerable time of assumed free-fall with all the mass at the central point. The shaded regions show a portion of the bulge before and after the collapse and it is clear that the bulge region has been stretched out.

The actual evolution of the protoplanet after leaving the perihelion region will contain elements of both the extreme conditions described above. There will be some partial recession of the tidal bulge towards the main body of the planet, returning it towards a spherical configuration. There will also be continued collapse and a stretching of the bulge material. An additional factor will be that all the mass is *not* concentrated at the central point and the form of collapse will be modified; in particular the tidal bulges may become less laterally extended due to inwardly acting gravitational forces within them. For the extended bulge material there may occur the conditions that Jeans postulated for the formation of condensations in a filament, and we shall consider the possibility of the formation of protosatellite condensations.

The analysis leading to the calculation of ω_f, the angular speed of the tidal tip, shows that the angular speed of bulge material further towards the centre is very little different, so that initially the bulge material is all rotating at about the same speed. If there was no transfer of angular momentum and the collapse of the bulge material was homologous, i.e. the distance each element moved towards the centre was proportional to its original distance from the centre, then all the bulge material would continue to rotate at the same speed. However, the stretching of the bulge implies that inner material moves inwards proportionately more than outer material and, by conservation of angular momentum, will progressively spin faster than outer material as time passes. The form of the bulge material at a late stage, when it has taken on a filamentary form, is shown in Fig. 8.3. It should be noted that the differential spin of the bulge material adds to the stretching-out process.

A protosatellite condensation forming in the filament would, at its time of origin, be moving in an eccentric orbit which would eventually round off in the resisting medium that would inevitably surround the newly formed planet. If the medium had a high central condensation then the final circular orbit would have a radius close to the original periastron distance.

This completes the general description of the satellite-forming process and we shall now examine our model in more detail to see whether or not this mechanism is plausible.

8.5 A MORE DETAILED STUDY OF THE MODEL

To be plausible the satellite-formation model should be able to show:

(1) that the filament contains sufficient mass to form the satellites,
(2) that the expected number of condensations within the filament should agree with the observed number of satellites,
(3) that the masses of individual condensations should exceed the Jeans critical mass and be consistent with the observed satellite masses,

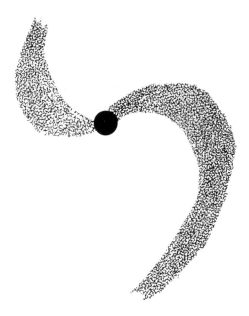

Fig. 8.3 — Form of the tidal bulges after the main core of the protoplanet has collapsed. The large bulge forms an extended filament-like structure.

(4) that the total angular momentum imparted to the tidal bulges should satisfy the requirements for both the planetary spin and the satellites in their orbits, and

(5) that the partitioning of angular momentum between planet and satellites should agree with observation.

Ideally, to show all the above points in an *ab initio* fashion would require the study of a model of exceptional complexity, which is possible in principle but which has not been done. Here we shall restrict ourselves to showing that the model is consistent, or at worst not inconsistent, with the requirements (1) to (5). In addition we shall mostly concentrate our attention on the major and very regular Galilean satellite system of Jupiter, although we shall also be referring to Saturn and Uranus in particular contexts.

(i)   The mass of the filament

The filament is assumed to be the source of satellite material, but in all our considerations so far we have been treating the protoplanet as consisting predominantly of molecular hydrogen. In fact the protoplanet consists of condensed interstellar material and will contain around 2% of solids in the form of grains of iron, silicates and ices, or perhaps even grains of mixed composition.

A protosatellite condensation will collapse if its mass is greater than a Jeans critical mass (equation (1.2)) for the density and temperature of the filament material. A rather crude, but not inappropriate, way of thinking about the Jeans mass is that it represents the condition where the *mean* speed of a gas molecule roughly equals the escape speed from the total mass of material. Where the escape speed is appreciably larger, then there will be collapse to high density on a fairly short timescale dependent on the kind of considerations which were dealt with in chapter 7. The dense grains will fall into the centre to form a solid core with the gas forming an envelope around it containing most of the mass. However, at this stage a new kind of stability must be considered — that of the gas envelope. The speeds of the gas molecules will have a Maxwellian distribution (Fig. 8.4) and those with the highest values, in the tail of the distribution, will be able to escape. This will occur on a much

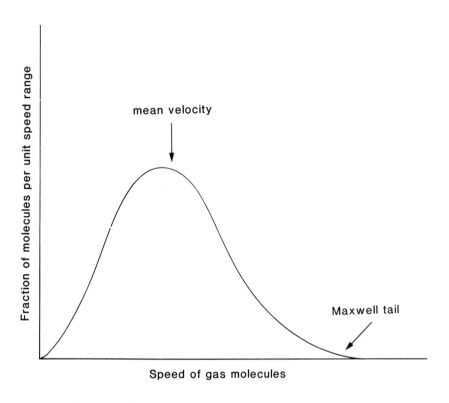

Fig. 8.4 — A Maxwell distribution for the speeds of gas molecules at a particular temperature.

longer timescale than the original collapse but inevitably, under the conditions of satellite formation, there will be left at the end just the solid cores with masses about 2% of the original masses. Of all the known satellites only Titan has been able to retain an appreciable atmosphere.

The total mass of the four Galilean satellites is 4×10^{23} kg and we must therefore expect that the total mass of the filament is at least of order 2×10^{25} kg (three times the mass of the earth). According to the model presented here, the major bulge, which contains more mass and intrinsic angular momentum than the smaller one, is to provide this material. It contains about 10% of the total volume of the condensation but we need to know its average density as well. The part of the original protoplanetary condensation, that within the radius R_c which is eventually to form the planet, will have a lesser central condensation of matter than is illustrated in Fig. 7.5 since the outer very diffuse regions are lost. A rough estimate, based on Fig. 7.5 is that the mean density in a tidal bulge region would be somewhat greater than 10% of the mean density. If this is so then the total mass of the bulge will be slightly more than 1% of the total planetary mass, or about 2×10^{25} kg. This agrees with expectation.

(ii)   The number of condensations in the filament

A schematic representation of the collapsed planetary core and the evolving filament at some late stage of the evolution is shown in Fig. 8.3. The exact time when a satellite came into being would be hard to define; as the filament develops, so gradually, concentrations of mass would form in it, initially connected to the remainder of the filament but eventually becoming distinct entities recognizable as individual protosatellites.

For the purpose of the present analysis we shall take a stage when the total length of the filament is D and we shall estimate the cross sectional area as $D^2/16$, giving a length : width ratio of about $4:1$. A higher ratio would be helpful to the process which is to be described, but we are taking this rather laterally diffuse example to illustrate the tolerance of the model to the arbitrarily chosen parameters. The mean density of the filament will be $16M_f/D^3$, where M_f is the mass of the filament and we shall take the filament as consisting of hydrogen with molecular mass m and with a temperature of θ. From equation (1.1) it can be shown that the length of a condensation will be

$$l = \left\{ \frac{k\theta}{16GM_f m} \right\}^{1/2} D^{3/2} \tag{8.7}$$

From this may be deduced the expected number of condensations

$$n_c = D/l = \left\{ \frac{16GM_f m}{k\theta} \right\}^{1/2} D^{-1/2} \quad . \tag{8.8}$$

For $M_r = 2 \times 10^{25}$, $m = 3.3 \times 10^{-27}$ (molecular hydrogen) and $\theta = 50$ K

$$n_c = 3.2 \times 10^5 \, D^{-1/2} \tag{8.9}$$

and for $n_c = 4$, the number of Galilean satellites, we find $D = 6.4 \times 10^9$ m. This is a very sensible value, being more than three times the orbital radius of Callisto, the outermost Galilean satellite, and we shall see later that such a value is to be expected.

This analysis has assumed a uniform filament, but in fact it would be expected that the width and density, and perhaps even the temperature, of the filament would vary along its length since the collapsing planet would be heating up through the dissipation of gravitational energy. Thus there would be variations of the length of condensations and the mass they contain; we have been dealing with an average situation.

(iii)   The mass of individual satellites
Based on the results previously found, the density of the filament material will be 1.2×10^{-3} kg m^{-3} and the temperature is assumed to be 50 K. The Jeans critical mass, as given by equation (1.2), is calculated as approximately 3×10^{24} kg, comfortably below the mass of each condensation (approximately 5×10^{24} kg). This indicates not only that the protosatellites would have collapsed, but that they would have done so quite quickly since gravitational forces were so much in the ascendent. The free-fall time for the estimated density is about 0.06 years, but the total actual collapse time would probably be of the order of a year, taking into account temperature-induced pressure effects during the collapse.

(iv)   Total angular momentum
To estimate the total angular momentum necessary to explain planetary spin and satellite orbits we must take into account that the satellites we see today are just the solid residues of bodies originally 50 times more massive. In Table 8.3 there is given the required initial angular momentum based on the assumption that the present satellites contain only two per cent of their original mass. The interesting conclusion is that for all three planets under consideration there was an approximate 2:1 partitioning of angular momentum between planetary spin and satellite orbits — suggestive of some relationship between the two rather than some haphazard and independent mechanisms.

In the first column there are the values of J_{2b}/f, coming from equation (8.6). If all the angular momentum generated in the evolving protoplanet was to be incorporated in planetary spin, as we see it today, and in augmented satellite orbits, then we can deduce values of f for Jupiter, Saturn and Uranus of 0.07, 0.08 and 0.02 respectively. The original elliptical satellite orbits would have rounded off in a resisting medium (see Chapter 9) so that the initial augmented angular momentum of the satellites would have been somewhat higher than that given in Table 8.3; this would increase the value of f by up to 20% but leave unchanged the conclusion that sufficient total angular momentum is available from the model with reasonable values of f, substantially less than unity.

(v)   Angular momentum of the satellites
From the value given for Jupiter for ω_r it can be found that the specific angular momentum (sam) of tidal-tip material after departure from the perihelion region is about 7.54×10^{13} m^2 s^{-1}. This must be compared with the sam for the orbit of Callisto, the outermost Galilean satellite, which is about 1.54×10^{13} m^2 s^{-1}. The fact that the value from the model is higher is encouraging since there are three factors which could reduce the sam of the tidal-tip figure down to the value for Callisto.

(a) The material forming Callisto will be contained in about one-quarter of the bulge material, most of it well away from the tidal tip. The sam of bulge material varies as d^2, where d is the distance from the centre of the protoplanet and the average sam of the outer 25% of the bulge material will be about 80% of that at the tip.

(b) By the time that the satellite forms in the filament, there would have been a considerable transfer of angular momentum inwards by viscous drag effects. This would have been greater for matter close in than for that further out.

(c) When the satellites form as compact bodies their initial orbits around Jupiter will be elliptical with a sam

$$Q_s = (GM_J p)^{1/2} \tag{8.10}$$

where p, the semi-latus rectum, is related to the semi-major axis and eccentricity of the orbit by equation (1.3)

$$p = a(1 - e^2).$$

The process of satellite formation would be bound to release material in the form of a resisting medium around the planet, and if this had a high central condensation then the orbit would round off to the perijove distance $a(1 - e)$ with sam

$$Q_p = \{GM_J a(1 - e)\}^{1/2} . \tag{8.11}$$

We find

$$Q_p/Q_s = (1 + e)^{-1/2} . \tag{8.12}$$

The length of filament previously calculated, its schematic appearance as shown in Fig. 8.3, and a perijove distance equal to the orbital radius of Callisto suggest $e \simeq 0.5$ or $Q_p/Q_s \simeq 0.8$.

Taking effects (a) and (c) together suggests that, without the mechanism (b) the sam of the outermost satellite would have been 4.8×10^{13} m^2 s^{-1}, or about three times as much as is observed for Callisto. Thus to obtain the correct sam for Callisto would require some two-thirds of the filament angular momentum to be transferred to the central body — a conclusion which agrees with the figures given in Table 8.3.

The above example relates to Jupiter but, as is shown in Table 8.4, the patterns for Saturn and Uranus are similar. The anomaly seems to be Uranus, where the value of f is much smaller than for the other two planets and suggests that the original quota of angular momentum was less than that indicated by the solar-tidal-effect model.

8.6   A SUMMARY

The description given here for the formation of satellites is restricted to showing the plausibility of the mechanism in broad terms. It can be fairly convincingly shown that the angular momentum acquired by the tip of a tidally distorted protoplanet is

Table 8.4 — Relationship of the specific angular momentum (sam) of the outermost regular satellite to the sam of the tidal tip as derived from the model

Planet	(a) sam of outer regular satellite	Initial sam of tidal tip	(b) Estimated sam of outer satellite without AM transfer	Ratio (a)/(b)
Jupiter	1.54×10^{13}	7.54×10^{13}	4.8×10^{13}	3.1
Saturn	6.81×10^{12}	2.36×10^{13}	1.5×10^{13}	2.2
Uranus	1.84×10^{12}	9.35×10^{12}	6.0×10^{12}	3.25

sufficient to explain the orbits of the outermost regular satellites, even allowing for loss processes. It would require more detailed modelling to show beyond doubt that the total angular momentum for both planetary spin and satellite orbits would be available, but the estimated small values of f are encouraging in this respect. The values of f also bear on the total mass of material available in the filament for satellite formation which the above analysis shows is only just enough for satellites, with nothing to spare.

A notable omission in our description has been consideration of a possible role for the smaller tidal bulges in satellite formation. While they would have had less mass and much less angular momentum than the main bulge, they were by no means negligible. Spacecraft observations have shown that all three of the planets we have considered here have ring systems and very small satellites closer to the planet than those observed by earth-based telescopes. It is not inconceivable that such close-in objects could have been derived from the lesser bulge.

The only outstanding anomaly in our results seems to be the value of f for Uranus, which is much smaller than for the other two planets and suggests that it had less original angular momentum than is indicated by the solar-tidal-effect model. Uranus is anomalous also in the direction of its spin axis, which is not very far from the plane of its orbit and gives a retrograde spin (see Table 2.3). It is impossible for the model we have been considering to explain this observation and, later, we shall make other proposals to explain the magnitude and direction of the angular momentum vector associated with Uranus.

9

Planetary orbits

9.1 INTRODUCTION

The Capture Theory predicts initial planetary orbits which are very eccentric compared with those observed today. Table 6.1 lists some orbits arising from the capture process and these are illustrated in Fig. 6.15. The smallest eccentricity is 0.73, compared to a value of 0.25 for the present-day Pluto. Indeed the eccentricities are more reminiscent of cometary than of planetary values. Although this may seem to be a serious fault of the theory, we shall see later that it makes some of the solar-system features often described as irregular much easier to explain than in a nebula type of theory. However, it is clear that we must describe a mechanism which would circularize or 'round-off' very eccentric orbits. Three possible mechanisms are available:

(a) tidal effects;
(b) gravitational point-mass effects;
(c) the action of a primitive resisting medium.

Although (a) and (b) are of great importance in our planetary system, we shall see that (c) is the only process capable of rounding the orbits.

9.2 TIDAL EFFECTS

The Capture Theory is highly dependent on tidal effects which cause the initial capture of planetary material. We have seen also how an extended body moving in a non-uniform gravitational field suffers a torque leading to a modification of its angular rotational speed (section 6.4). From a consideration of energy and angular momentum, which must be conserved, we can show that the orbital elements must also change.

The total energy of an orbiting planet can be split into two parts: the rotational energy ($E_S>0$) and the orbital energy ($E_O<0$, for an elliptic orbit). Since energy must be conserved in a non-dissipative system we can write

$$E_S + E_O = E \text{ , a constant.}$$

Similarly the total angular momentum, H, which is also a constant, can be split into orbital and spin components:

$$H_S + H_O = H \text{ , a constant.}$$

Any change in the period of rotation yields changes in E_S and H_S, and so the orbital components, E_O and H_O, must change by the same amounts (and with opposite sign) to maintain the constant values E and H.

Consider now a protoplanet being formed on an eccentric orbit and with zero spin. The orbital motion in the non-uniform gravitational field will result in a direct spin approximately equal to the perihelion rate. This will be achieved very rapidly if the protoplanet is very extended. if the protoplanet collapses very quickly to a radius very small compared with the dimensions of the orbit, the spin will increase accordingly but, being the result of collapse, this will not remove angular momentum from the orbit. The orbit will be changed only as a result of a gravitational coupling between the body rotation and the solar field. In Fig. 9.1 we present a graph showing

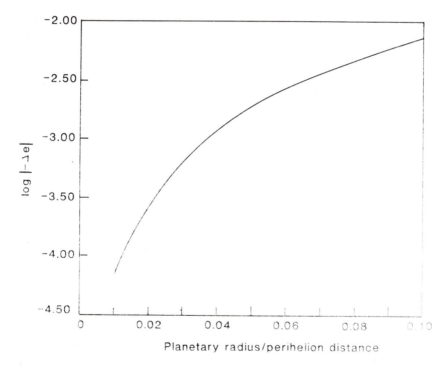

Fig. 9.1 — Change in eccentricity when protoplanet is spun up to perihelion rate.

the reduction in orbital eccentricity due to the planet being spun up to the perihelion rate for a range of (constant) radii. Even when the planet remains at a radius one-tenth of the perihelion distance, the eccentricity is reduced by no more than 0.0072 from an initial 0.9. Consequently this mechanism cannot produce the required rounding effect.

Only the three innermost planets are affected significantly by tidal forces over the lifetime of the solar system. In the case of the earth, spinning much faster than the orbital rate, the solar tide tends to reduce the angular velocity. This is balanced by increases in the orbital angular momentum and energy, and consequently by a higher eccentricity. However, the effect is very small and in any case complicated by the lunar tides which are about twice as great. Solar tides completely dominate the rotation of Mercury, while that of Venus, complicated by the massive atmosphere, is much slower than for any other planet. It seems possible that Venus's rotation is affected by tides due to the earth, although there is some controversy over this.

Since the planets would probably condense fairly quickly, any angular perturbations due to tides in the major planets would be short-lived. However, there would remain the dissipation of energy due to variations in tide heights during an eccentric orbit. A planet would suffer the greatest tidal distortion near perihelion and least deformation would occur near aphelion. During each orbit the planet would undergo the complete cycle of deformations, with concomitant variation in its potential energy. Frictional effects would cause some small fraction of this energy change to be converted to heat, just as the bending to and fro of a metal bar causes heating. If we assume no angular perturbations, the orbital angular momentum must stay constant while, at the same time, friction is removing energy from the orbit. Under these conditions the theory of orbits tells us that the orbit should be rounded eventually to a radius equal to the initial semi-latus rectum. If a and e are the semi-major axis (mean radius) and eccentricity of the orbit of a planet of mass m, and $M_\odot$ is the solar mass, then the energy and angular momentum (per unit mass) of the planet are given by

$$E = -G(M_\odot+m)/2a. \tag{9.1}$$

and $\quad H = \sqrt{\{G(M_\odot+m)a(1-e^2)\}} \tag{9.2}$

Thus it is clear that dissipation of energy (E is reduced), combined with conservation of angular momentum H, causes a decrease in the semi-major axis while keeping the semi-latus rectum $p=a(1-e^2)$ constant. This requires a reduction in the eccentricity e. The end result must be an orbit with $e=0$ (circular) and radius p. Of course the dissipation of energy will cease when the orbit becomes circular because the tide height is then constant.

This process can be completed only if there is sufficient time. In 1974 Dormand and Woolfson showed that the timescale was dependent on the planetary radius, which would not remain sufficiently large to give near-circular orbits. To get some idea of the magnitude of the effect, one can compute the timescale for Mercury, which is currently the only planet known to be completely under solar tidal influence. With the assumption that one per cent of the change in gravitational potential energy due to variation in shape is dissipated (converted to heat) per orbit it turns out that the rounding time is 6×10^9 years. This result is hypothetical because the angular tidal effects are certainly not negligible in Mercury's case.

9.3   GRAVITATIONAL POINT-MASS EFFECTS

A planetary system with the initial state predicted by the Capture Theory would probably undergo substantial change as a direct result of gravitational forces between planets. The major planets (Jupiter in particular) would cause large perturbations in the orbits of any nearby bodies, but such perturbations may not tend to reduce eccentricities. A close encounter between planets could either increase or reduce eccentricities depending on the parameters of the interaction.

The effects of perturbing forces in the solar system have been investigated by a number of different methods, all of which contain some element of approximation since methods of mathematical analysis do not permit precise predictions if more than two bodies are present. When short timescales are involved, predictions of the motions of the planets can be made fairly easily. For long periods of time, prediction is much harder, although the introduction of powerful computers and sophisticated numerical techniques have led recently to great advances, with up to 10 million orbital periods being analysed. Thus it is possible for the outer solar system to be investigated over a period of 100 million years — a non-negligible fraction of the age of the planets. Such calculations show little or no change in the state of the system; no systematic increase or decrease in orbital eccentricities occurs.

If rounding-off had to be a result of gravitational interactions between planets then a reasonable extrapolation of these results would lead to the prediction that the solar system had existed for a period much greater than the supposed value of about 5×10^9 years. However, what the calculations tell us is that under gravitational forces *alone* there appear to be no substantial changes taking place over very long periods. The implication is that other physical processes were active in the remote past.

9.4   A RESISTING MEDIUM

The Capture Theory model predicts the capture of much more material than would be necessary to construct the planets. The simulations described earlier indicate that much of the material would undergo further dispersion. We have suggested (Dormand and Woolfson, 1974) that this material would form a resisting medium, itself orbiting the sun, which would affect the motions of the condensed bodies. With a small understanding of celestial mechanics it becomes very easy to see that such a resisting medium would be capable of rounding off a planetary orbit.

Assume that the medium consists of a large number of small particles, each one pursuing a circular heliocentric orbit. Since a particle moves principally according to solar gravity (although very small particles are influenced by radiation which reduces the effective central mass) we can easily calculate its velocity for any orbital radius. This is given by:

$$v^2 = \mu/r .$$
(9.3)

where v is velocity, r is the orbital radius and $\mu = GM_\odot$. Now suppose a planet has a perihelion distance r and orbital eccentricity e; its velocity V at perihelion satisfies

$$V^2 = \mu(1+e)/r .$$
(9.4)

and thus it is evident that $V > v$, and that the planet is moving faster than the medium

(Fig. 9.2). Since some of the particles of the medium will impact on the planet it is clear that the resulting force will cause deceleration.

A planet with aphelion r has velocity U satisfying

$$U^2 = \mu(1-e)/r \tag{9.5}$$

and thus moves more slowly than the medium at that distance. At aphelion, therefore, the medium will tend to accelerate a planet. The formulae for U and V above allow us to determine the effect of deceleration and acceleration at the extremes of the orbit of a planet. If we fix the value of r and reduce V, then e must decrease; similarly increasing U will reduce e. The eccentricity e is consequently reduced by the resisting force at both extremes of the orbit (Fig. 9.3) when the planet moves within a freely rotating medium, which therefore becomes a possible mechanism for the rounding-off process.

For the mechanism to be plausible we must show that the eccentricity of the orbit of a planet can be reduced to a small value, less than 0.1 say, in a reasonable time. This means that the actual decelerating force on a planet at any time due to a freely rotating medium must be calculated.

First we consider the effect of the impact of the medium on a planet passing through it. Suppose the planet has radius R and has velocity V with respect to the medium (Fig. 9.4(a)). In every unit time a cylindrical volume of 'medium' (radius R, length V) will be collected by the planet. Thus the momentum change per unit time (i.e. force) is the product of the mass and velocity:

$$F = \pi R^2 p V^2. \tag{9.6}$$

Actually we have underestimated considerably the force F since the gravitational effect of the planet has been neglected. This would cause particles of the medium to be deflected towards a planet rather than streaming past in parallel fashion. Consequently the radius x of the 'tube' of material accreted by the planet is increased considerably (Fig. 9.4(b)), particularly when the planet is very massive. This is an example of the Eddington accretion mechanism illustrated in Fig. 5.16. Fig. 9.5 shows the variation of 'gravitational' radius x with planetary mass and relative velocity. It will be clear, of course, that the relative velocity of the planet would vary around the orbit and would have opposite signs at the two extremes.

In fact we have still not considered the total forces acting on a protoplanet. All particles out to distance x from the centre of the planet will be accreted (Fig. 9.4(b)) but particles much further away will be deflected. The changes in their momenta can be calculated, yielding an increased force serving to modify the planetary orbit. The limiting distance for affected particles is known as the sphere of influence of the planet

$$S = r(M_p/2M_\odot)^{1/3}, \tag{9.7}$$

where r is the distance from the sun. This is so much larger than the gravitational radius that the latter becomes almost negligible in the resisting medium calculation. To take an example, we consider the planet Jupiter at a perihelion distance of 5 AU.

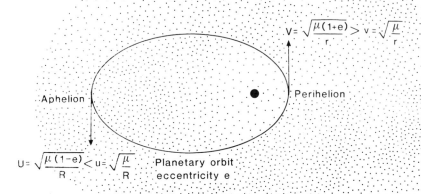

Fig. 9.2 — Differential velocities. At perihelion planet moves faster than resisting medium, at aphelion planet moves more slowly than the medium.

If its eccentricity were 0.8, the sphere of influence, S, would be approximately 64 times as large as the radius x.

Formula (9.7) also implies that the resisting medium will be more effective at large heliocentric distance. For present-day Jupiter the sphere of influence, S, is about 0.4 AU, for the earth 1/90 AU and for Neptune 0.9 AU. Thus we might expect the rounding-off process to be relatively efficient in the outer solar system.

It is unlikely that newly formed planets would move in a medium of constant density. From the results of Chapter 6 we might expect the medium to have greatest density at a radius of around 5 AU, the orbital radius of Jupiter. Also the initial mass of the medium would be greater than the combined planetary masses. The capture event models do not predict precisely such values, but from present-day parameters in combination with simulation results we would estimate a total mass of 5 Jupiter masses for the medium with a maximum density at 5 AU (see Fig. 9.6). Of course the medium itself will evolve due to the actions of the planets. Some material will be accreted and absorbed by planets. To test the sensitivity of the rounding process with respect to changes in mass distribution, we also computed evolutionary sequences for a medium with a simple exponential distribution (Fig. 9.6).

Once the relevant perturbing forces have been determined, it is possible to compute motions of planets within a resisting medium. In this instance our requirements are a bit different to those of Chapters 6 and 7. We do not want to know precisely where a planet is at some instant of time; it is sufficient to determine the variation with time of the orbital elements of the planet. In fact we require only the

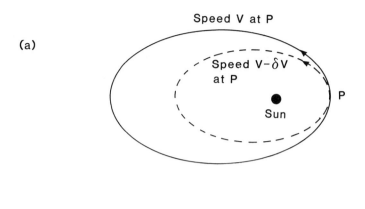

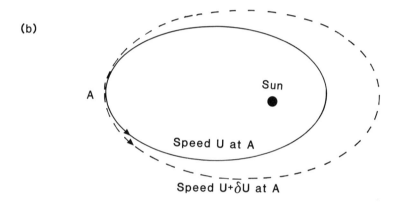

Fig. 9.3 — Effect of velocity differential: (a) at perihelion; (b) at aphelion.

values of eccentricity and semi-major axis, and these are determined by the energy and magnitude of the angular momentum. Thus we actually compute the rates of change of these quantities. The calculation of such an evolutionary sequence is much less demanding computationally than the tidal disruptions considered earlier and it is fairly easy to deal with time durations of the order of the age of the solar system.

Unlike the tidal forces encountered earlier, the resisting force of our medium is very small indeed, and to a first approximation the orbit of a planet is not affected at all. Consequently we can obtain by mathematical methods a formula for the change in orbital eccentricity, e, resulting from a single orbit within the medium. This formula is fairly complicated but is made much less so by our justifiable assumption that a and e remain virtually constant over one orbital period. Suppose Δe is the change in eccentricity over a single period $T = 2\pi\sqrt{(a^3/\mu)}$. The quotient $\Delta e/T$ represents an average rate of change of eccentricity and may be used (in conjunction with $\Delta a/T$) to determine the evolutionary profile of the planetary orbit.

Fig. 9.7 shows graphs of the variation of the eccentricity and semi-major axis of

(a)   Planetary gravity ignored

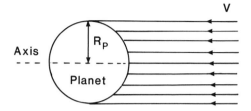

Material up to distance R_P from axis is accreted.

(b)   Planet attracts medium

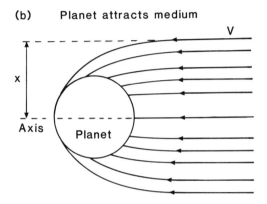

Material up to distance x from axis is accreted.

Fig. 9.4 — Accretion: (a) planetary gravity ignored; (b) planet attracts medium.

the orbit of a planet of half the mass of Jupiter moving in resisting media of the types described above. It is clear that in each case a significant rounding occurs within a relatively short period of time. Also the two profiles are very similar in appearance, indicating a lack of sensitivity of the mechanism with respect to the distribution of mass. We note that the semi-major axis decreases in each of these cases.

The resisting medium will always tend to round off orbits but the semi-major axis will not necessarily be reduced. In Fig. 9.8 we show the variation of rounding times and final *a* with initial semi-major axis. Orbits with small semi-major axes (in relation to the position of maximum density of medium) tend to gain energy, while the larger orbits tend to lose it. This is not too surprising if one considers the rounding effects at the extremities of an orbit. The 'small' orbit will suffer the largest perturbation near aphelion if the density there is highest. Consequently the semi-major axis will increase. The effect is reversed for larger orbits.

The actual rounding time will depend on the planetary mass and also the total amount of medium material. The variation of the former is illustrated in Fig. 9.9; a planet such as Jupiter will round off in about 100 000 years, but the earth (in the same orbit) would take 1000 times as long. We would expect a linear variation of rounding

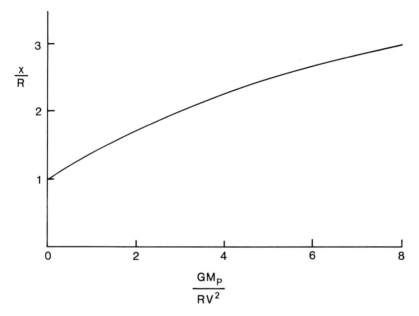

Fig. 9.5 — Gravitational radius of a planet.

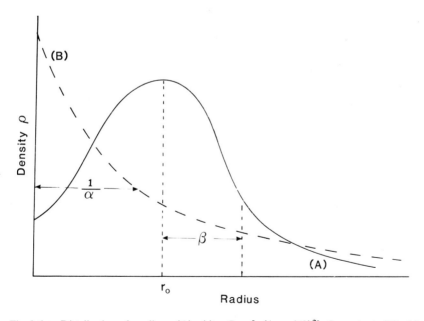

Fig. 9.6 — Distribution of medium. (A) $\rho(r) = C \exp[-\{(r-r_0)/\beta\}^2]$, C constant; (B) $\rho(r) = D \exp[-\alpha r]$, D constant.

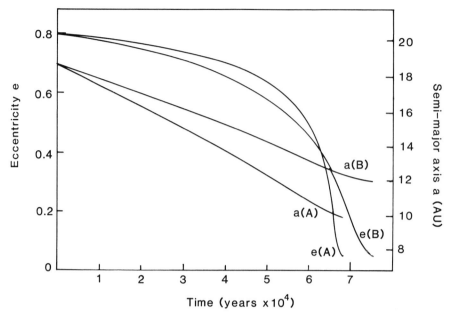

Fig. 9.7 — Orbital evolution in a resisting medium. Total mass of medium $= M_\odot/200$; Planetary mass = one-half Jupiter's mass; $\beta = r_O = 1/\alpha = 6.67$ AU; (A) indicates Gaussian distribution; (B) indicates exponential distribution (Fig. 9.6).

time with reciprocal mass of medium; i.e. if the medium mass is halved then the rounding time will be doubled.

So far, the evolution of the resisting medium itself has not been considered and the results referenced above have been computed under the assumption that no changes in the medium occur. This cannot be strictly true since at the present time the solar system does not contain our proposed medium. For our hypothesis to merit serious consideration we must explain the disappearance of material of mass greater than any of the known planets. Fortunately there are some well-established mechanisms which would cause this to happen.

The medium would certainly be modified by the 'stirring' effect of the planets. If most planets possessed greater semi-major axes than the maximum density radius of the medium then the reduction in their energies would tend to increase the energy of the medium; this would act in the same direction as solar heating. Energy from the latter would cause the gaseous component of the medium to be lost gradually. This would be a simple case of atoms and molecules with velocities greater than needed for escape passing into interstellar space. Hoyle (1960) has estimated that a significant amount of a gaseous nebula would be lost in this way over a period of about 3×10^7 years. This is much greater than the rounding time for major planets and so we consider that the previously described results are valid for these bodies. Yet it seems that the rounding of the orbits of terrestrial planets may not have been achieved in the lifetime of the gaseous part of the medium.

The solid component of the resisting medium (about one percent) would also be

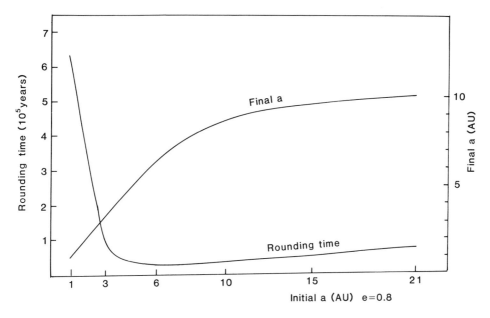

Fig. 9.8 — Variation of rounding-off time and final semi-major axis with the initial semi-major axis. Initial eccentricity is 0.8 and medium parameters are as in Fig. 9.7(A).

affected by solar radiation. Very small particles would be driven outwards by radiation pressure, but others would move inwards. Any particle orbiting the sun will receive radiated energy. When this radiation is re-emitted by the particle it will have a component of angular momentum with respect to the sun. The particle loses angular momentum and the orbit gradually decays, eventually intersecting the sun. This mechanism is known as the Poynting–Robertson effect (section 3.3, Fig. 3.5). We have shown (Dormand and Woolfson, 1971) that it would take at least 10^9 years for the solid component of the medium to be removed in this way. Thus it is likely that the resisting medium would be capable of rounding off the orbits of all the planets.

9.5   CONCLUDING REMARKS

We have shown, in this chapter, that a planetary system with primitive orbits of high eccentricity can evolve into a system like our own, possessing, in the main, near-circular orbits. Given the capture mechanism for the acquisition of planetary material, there can be little doubt that a resisting medium would be present in the early solar system. Thus the rounding mechanism is a natural consequence of the capture hypothesis.

   It is interesting to recall the discussion of point-mass effects (section 9.3). In the not too distant future it will be possible to extend such calculations to cover the lifetime of the solar system. We predict that these will 'reveal' further the long-term

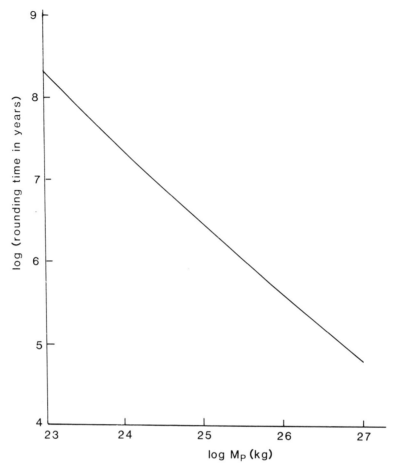

Fig. 9.9 — Variation of rounding-off time with planetary mass. Medium as in Fig. 9.8 and
initial orbit $(a, e) = (16.67 \text{ AU}, 0.8)$.

stability of the system. However, they will not reveal the 'origin' of the system
because evolutionary factors (such as a resisting medium) have long since disappeared and it is really not possible to cater for unknown features in any backward
simulation. To make progress in the field of cosmogony, hypotheses must be
formulated and tested in the manner suggested in this work. Sometimes such testing
leads to further insights into the problem. We would claim that the sequences of
events described above are logical developments of the capture hypothesis. Similarly
a planetary collision, which is the subject of the next chapter, falls into the overall
pattern of the Capture Theory.

10

A planetary collision

10.1 INTRODUCTION

We have seen in Chapter 6 how the early planetary orbits would have been very eccentric compared with the present ones. The high eccentricities turned out to be an advantage from the point of view of planetary condensation, which required a reduced tidal field for the major part of the collapse to take place. The rounding effects of uncondensed material were considered in the last chapter, although the treatment was restricted to two dimensions; this is not a disadvantage when the rounding-off times are to be determined, but we shall see later that non-zero inclinations have some significance in other aspects of early orbital evolution.

We have restricted our attention to the origin of major planets in our description of both the capture mechanism and planetary condensation. Certainly the densities found in early captured material would not support direct formation of planets of terrestrial mass. In an objective sense the inner planets are insignificant, accounting for about one-half per cent of the total planetary mass, but we do have a special interest in at least one of them! Since they are so different to the major planets, it may not be unreasonable to consider a different mode of formation. In this chapter we shall consider a planetary collision which causes the break-up of one of the bodies; some of the larger fragments can then form terrestrial planets. Actually this idea is not so different to the basic idea of capture which proposes that planetary formation is a result of the disruption of a larger body.

A glance at the early planetary orbits (e.g. Fig. 6.15) suggests the possibility of significant interactions since there appear to be intersections. The rounding process will reduce such possible occurrences in time, but we shall see that the characteristic timescale for major perturbations through planetary interactions are comparatively short. Therefore it is plausible that such processes have played an important part in shaping the solar system.

10.2 THE FREQUENCY OF PLANETARY COLLISIONS

We refer again to a typical system of early orbits shown in Fig. 6.15; these do not exhibit many intersections, but we shall see later that evolutionary processes would increase the number of possibilities. How can we determine whether or not two planets in intersecting elliptical orbits will collide or suffer a major perturbation through the gravitational effects of a close encounter? There are two approaches: the first assumes that we know accurately the initial conditions, while the second is statistical. We consider the deterministic case first.

In Fig. 10.1 we show two orbits intersecting at points O and P; it is assumed that

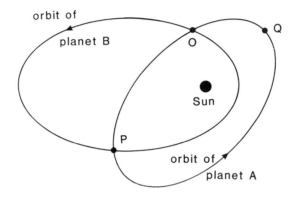

Fig. 10.1 — Intersecting orbits. Collisions or close encounters possible only near points O and P.

they are coplanar. Assume planet B (with the more eccentric orbit) is at O initially, and the other (A) is at Q. Suppose planet A takes time τ to reach point O (by which time B will have moved away). If A and B have orbital periods T_A and T_B then they will reach point O simultaneously if the equation

$$\tau + mT_A = nT_B$$

is satisfied for some integer values of m and n. (B is at point O at times O, T_B, $2T_B$, etc., and A reaches the same point at times τ, $\tau + T_A$, $\tau + 2T_A$, etc.) For a collision the equation does not have to be satisfied exactly since the collisional cross-section for the planets will be enlarged by their mutual gravitational attraction (see section 9.4). A close encounter or near miss would be less catastrophic, but could still cause major changes in the orbital elements of the planets. For such an event the equation above needs to be satisfied only approximately. Given a set of precise initial conditions, the value of τ could be determined and consequently we could find the values of n and m required for various encounter parameters. As hinted earlier, the situation is more complicated than this because the orbits evolve in shape and *orientation* due to resisting-medium effects. Consequently the points of intersection of the orbits will vary, thus invalidating any results predicted from the above analysis. In any case we do not have precise initial conditions for the solar system and so we cannot properly apply the method for this reason.

We are thus led towards a statistical analysis of the collision/close encounter scenario. Here we can assume that initial positions are distributed at random and consider the *probabilities* of the events of interest. It will be convenient still to determine approximate 'timescales' for these events. We consider again the two orbits depicted in Fig. 10.1. Let us select a particular time when planet B is at point O; it will be in this same position one orbit later and so on (assuming fixed orbits for the moment). We must find the probablity that, at the same time, planet A is close enough to O for a significant perturbational event. An actual collision is, of course, classed as a major event and the conditions required for this will be included within those considered here.

The term 'probability' is in common usage but some explanation is desirable here. The scale of probabilities is defined so that an event which is certain to occur has unit probability; an event certain *not* to occur has probability zero. For example the probability of achieving a 'head' with a single toss of a coin is one half; a 'tail' has the same probability. This value can be related to *expected* values. When a coin is tossed 100 times (say) we would 'expect' 50 heads, although an actual experiment may yield a slightly different value.

Application of statistics to orbital collisions or other major events is much more complicated than the coin-tossing example, but by making a number of simplifying assumptions we can find an estimate of the expected time which will elapse before a major event, involving two particular bodies, takes place. Thus we can say that the time spent by planet A in the critical region of its orbit (arc length *s* in Fig. 10.2) is

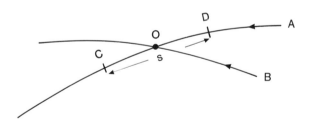

Fig. 10.2 — Intersecting orbits. When planet B is at O, planet A will suffer a major perturbation if it lies between C and D.

given by

$$\tau = s/v \qquad\qquad (10.1)$$

where *v* is the orbital speed in that part of the orbit (although the orbital speed varies continuously, we are justified in considering an 'average' value here). If the orbital period of the planet is T_A then the probability of it being in that region at any particular instant is

$$P_c = \frac{\tau}{T_A} = \frac{s}{vT_A} . \tag{10.2}$$

On average, planet B will have to cross the orbital intersection point $1/P_c$ times before it finds planet A in the critical region. This corresponds to an average time for an interaction

$$t_I = \frac{vT_A T_B}{s} . \tag{10.3}$$

where T_B is the orbital period of planet B. The time t_I is called the 'characteristic time' for the event in question.

In Table 10.1 is given a set of early planetary orbits which we believe to be

Table 10.1 — Characteristics of the early solar system

Planet	Mass $(\times M_\oplus)$	Radius $(1000\,\text{km})$	SMA(^{a}) (AU)	Period (years)	Ecc. e	SLR (AU)	Incl. i	Rounding time (years)
Neptune	18	28	62.3	492	0.720	30.2	3°	2×10^6
Uranus	15	26	35.6	212	0.690	18.8	2.5°	2×10^6
Saturn	100	66	18.6	80	0.680	10.1	1.5°	3×10^5
Jupiter	330	78	14.6	56	0.800	5.3	2°	1×10^5
A	17	27	12.2	43	0.874	2.9	1°	2×10^6
B	5.5	21	9.1	27	0.908	1.6	1°	6×10^6

SMA Semi-major axis.
SLR Semi-latus rectum.

consistent with a Capture Theory origin; the values or orbital inclinations will be discussed in section 10.3.

Let us consider the characteristic timescale for planets A and B. The only quantity in the formula (10.3) which is not directly available is the orbital speed v of the planet A in the critical region. The upper limit for v is around 33 km s^{-1} at the perihelion distance; the lowest speed possible will be that which applies when the intersection is at the aphelion of B (17.3 AU) where v is 5.5 km s^{-1}. Taking $v = 12$ km s^{-1} will lead to an error of not more than a factor of three, but probably much smaller. With $s = 10^5$ km, corresponding to a grazing encounter if a reasonable deflection is assumed (Fig. 9.5), equation (10.3) now yields 4.4 million years. A more realistic value for t_I would be half this, since there are two points of intersection near which an event is possible. Since the characteristic timescale falls between the round-off times for the two planets a close encounter between them looks quite plausible.

Applying the same formula to Jupiter and Saturn yields a smaller timescale but, since they have comparatively short rounding times, a major interaction between these planets does not seem very likely.

Of course the real situation would be much more complicated than the approximate picture presented so far. The rounding process, due to the resisting medium, continuously modifies the early planetary orbits and so the probability of a collision per orbit is always changing. Indeed it seems reasonable to predict that, for at least a part of the time of rounding, a pair of orbits may not intersect at all.

10.3   NON-COPLANAR ORBITAL EVOLUTION

To appreciate the non-coplanarity of the solar system it is necessary to consider the three-dimensional specification of an orbit. The size and shape of an elliptic orbit are uniquely defined by its semi-major axis and eccentricity. The orientation of the orbit is specified in terms of three angles which are illustrated in Fig. 10.3. These angles

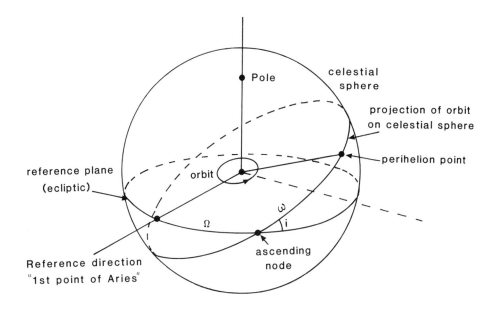

Fig. 10.3 — The orbit in space. Angles: i is the inclination, Ω is the longitude of the ascending node, and ω is the argument of perihelion.

are: inclination (i), longitude of the ascending node (Ω), and argument of perihelion (ω). For the solar system we normally use the ecliptic plane (orbital plane of the earth) to define inclinations. The 'X-axis' is the direction of the sun when it crosses the earth's equator moving north; this is called the vernal equinox. A planet with an inclined orbit will cross the equator at a different point, and the discrepancy between this and the vernal equinox is called the longitude of the ascending node. The argument of perihelion is measured from the planetary node.

The capture mechanism would certainly yield *near*-coplanar orbits. This is in line with the observed planetary orbits which have very small inclinations (Pluto is exceptional with $i = 17°$). However, any oblique protostar rotation would result in variations in inclination for the captured material. In 1977 we computed an upper limit of 9.2° for planetary inclinations with respect to the orbit plane of the protostar. This would be attained when filament material comes from the protostar when it is just rotationally unstable with an axis in its orbit plane, i.e. its equator would be perpendicular to its orbit plane in a similar fashion to that of Uranus, and material would be thrown out of the sun–protostar orbital plane. The inclinations presented in Table 10.1 reflect this result. What would be the effect of varying inclinations on the evolving planetary orbits? The major influence on early orbits would be the resisting medium which was formed from uncondensed material. This material would be mainly gaseous, but a solid component, in the form of very small particles or dust, would account for about 2% of the total mass of the medium. Now a solid particle in heliocentric orbit must move in a plane containing the sun. Gas molecules would not be thus constrained since they could be supported by the gas pressure gradient perpendicular to the mean plane. Hence we would expect the gaseous part of the medium to have a considerable vertical dimension. Since all the solid particles must pass through the mean plane they suffer collisions near ascending or descending nodes. Such collisions would tend to destroy vertical velocity components and eventually bring all the dust particles into the mean plane. Thus a very thin disk of dust material, perhaps only a few centimetres thick according to Lyttleton (1972), would form. This disk may have been similar to the rings of Saturn which are extremely thin — probably for the same reason. This disk would not have been important in the orbital rounding process, but, as we shall see, it has another effect of significance with regard to enabling strong interactions to take place.

In our previous analysis of the rounding process we neglected the gravitational field of the resisting medium itself. The field of the gaseous component being very extended, would not produce any qualitative difference in the rounding effect, but the field of the dust disk would certainly introduce small non-central forces (i.e. forces not pointing toward the sun) which particularly affect planets with inclined orbits. Such forces tend to affect the orbital planes themselves and therefore result in changes to the orientation angles ω and Ω. These changes take the form of rotation of the perihelion longitudes ($\omega + \Omega$) of the early planets. This type of motion (precession) is shown in Fig. 10.4; the rate of precession has been exaggerated to show the effect more clearly. A body moving within the plane of the disk would experience only central forces, but its orbit still would suffer changes in the apse line; this would be due to the variation with orbital radius of the effective mass of the central body. The type of precession being considered here is very closely related to that exhibited by satellites (natural and artificial) orbiting oblate planets; the oblateness factor of a planet can be interpreted in terms of an equivalent disk component.

The precession of orbits will certainly cause apparent points of intersection of pairs of orbits to vary. In fact inclined orbits are unlikely to intersect precisely in space, although they may do so in projection. The best-known example of this is the case of Pluto and Neptune. Many illustrations of the orbits of these two planets apparently indicate intersections, but such diagrams show only the projected ellipses. The number of possible close encounters, including collisions, is greatly increased by precession because it ensures that most of the orbits listed in Table 10.1 will 'intersect' at some time or other. Even if the orbit of planet A were to lie entirely within that of Jupiter at the time of formation, precession would eventually cause a

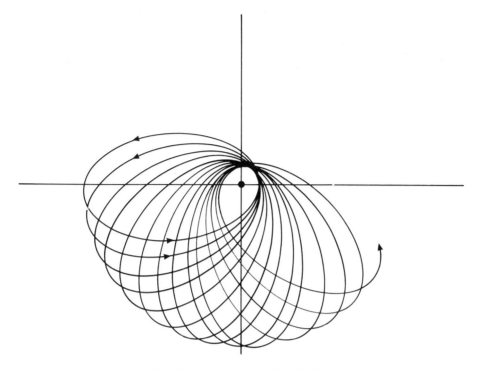

Fig. 10.4 — A precessing elliptical orbit.

near intersection and hence a possible collision, provided that the two precession rates were different.

In 1977 we computed some evolutionary sequences for planets with small orbital inclinations affected by the resisting medium, as discussed in Chapter 9, and also by the disk force. The variation of the orbit of a planet of mass half that of Jupiter with initial elements $(a, e, i, \omega, \Omega) = (33.3 \text{ AU}, 0.9, 2°, 0°, 0°)$ is shown in Figs 10.5 and 10.6. These results were computed in a different way to those of the last chapter and, although they demonstrate the same times for round-off, there are some new features. Comparing Figs 10.5 and 9.6 we note an obvious change in the initial behaviour of the eccentricity; with the new method we see a fairly rapid initial decrease. This discrepancy is due to the assumed initial conditions which were defined in terms of a Keplerian elliptic orbit with elements given above. Because of the relatively large perturbation of the resisting medium, the standard orbit is not a particularly good approximation of the actual motion. In other words there is a non-negligible change in the orbit elements during a single orbit. This effect is not a feature of earlier results which were computed differently making use of the assumption that elements undergo negligible change during a single orbital period. The final state of an orbit is very nearly the same with both methods.

The evolution of a and e is now accompanied by variations in i, ω, and Ω. The

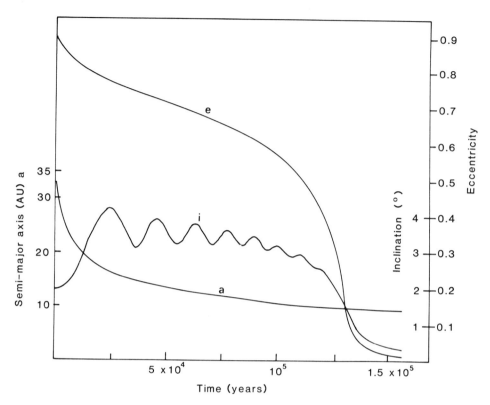

Fig. 10.5 — Orbital rounding-off with initial values (a, e, i) = (33 AU, 0.9, 2°). Medium parameters as in Fig. 9.7 with a Gaussian distribution.

inclination shows some periodicity in its variation, but a considerable reduction is evident. In Fig. 10.6 we see that the other angular elements circulate, thus indicating a precession as illustrated if Fig. 10.4. The value of the precession is about 1° per orbit. This will vary according to the elements of the orbit, in particular the values of a and e. It may be shown that the rate of precession is inversely proportional to the square of the semi-latus rectum of an orbit and that of the node is approximately proportional to the cosine of the inclination when it is greater than zero (the node is indeterminate when i is zero). Since all the inclinations are small in our example, the variations in semi-latera recta will be most significant in differentiating precessional rates and we therefore expect planet B to precess about four times as fast as Saturn (Table 10.1). The differential apse rotations are most important since uniform rotation would not increase the number of major-interaction possibilities.

10.4   ENCOUNTERS IN AN EVOLVING SYSTEM

Taking into account the orbital evolution described above and the probability considerations of section 10.2, we have computed various characteristic times for

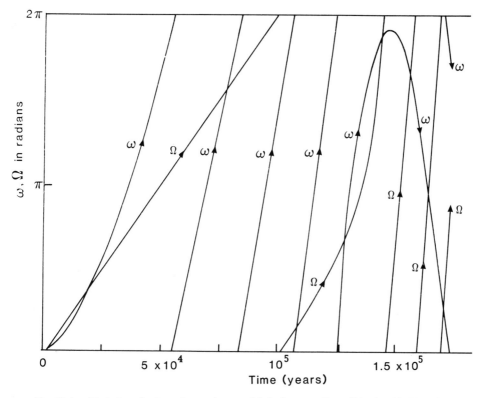

Fig. 10.6 — Variation of orientation angles ω and Ω during rounding-off depicted in Fig. 10.5.

major events in the early solar system with characteristics depicted in Table 10.1. Three types of event are classified as being 'major'; these are:
(a) encounters after which the planet with the smaller semi-major axis leaves the solar system;
(b) encounters after which the planet with the larger semi-major axis leaves the solar system;
(c) collisions.

The first two categories require some explanation. Following any encounter between two planets their orbital elements will be changed; the closer the encounter (minimum separation), the greater will be the changes (Fig. 10.7). A practical example of this type of effect is provided by the space probe Voyager I. This was placed in an elliptical transfer orbit between the earth and Jupiter, with aphelion only just beyond the Jovian orbit. A close approach or encounter with Jupiter was used to perturb the probe's orbit so that the new aphelion exceeded that of Saturn. Obviously the event was timed so that Saturn was in the best place at the time its orbit was crossed and so a second perturbation of major proportions occurred. This transformed the probe orbit to hyperbolic type and so it has, in effect, been ejected from the solar system. It is reasonably straightforward to compute the parameters required for this type of event since one of the bodies involved is of negligible mass

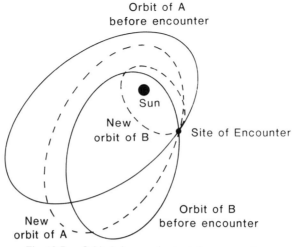

Orbit of A
before encounter

Sun

New
orbit of B

Site of Encounter

Orbit of B
before encounter

New
orbit of A

Fig. 10.7 — Orbital changes due to a close encounter.

and so we do not have to consider its effect on the planet and the sun. When both
interacting bodies are massive then it is possible for either of them to be ejected from
the system. It is more likely that the less massive body will be ejected, but for
moderate mass ratios the larger body could be lost instead. Using the ideas of section
10.2 we have computed characteristic times for interactions of the three types (a),
(b), and (c) (Dormand and Woolfson, 1977) and these are given in Table 10.2. A
fourth characteristic time, that for *any* one of the three, is also tabulated.

Table 10.2 — Characteristic times for events in the early solar system

Planets		Time in millions of years			
Inner	Outer	(a)	(b)	(c)	(a, b or c)
B	A	2.41	1790	33.3	2.24
B	Jupiter	0.11	∞	13.1	0.11
B	Saturn	2.53	∞	41.2	2.38
A	Jupiter	0.09	∞	5.91	0.09
A	Saturn	2.56	339	28.1	2.33
A	Uranus	∞	183	465	131
Jupiter	Saturn	453	0.22	139	0.21
Jupiter	Uranus	∞	0.34	111	0.34
Jupiter	Neptune	∞	0.95	327	0.94
Saturn	Uranus	∞	4.40	314	4.32
Saturn	Neptune	∞	17.8	1150	17.5
Uranus	Neptune	3420	2190	5240	1060

The times given in Table 10.2 are computed on the assumption that the
eccentricities remain as in Table 10.1; only the directions of the apse lines were

allowed to vary. The most likely major events seem to involve planets A, B, and Jupiter. Indeed all interactions with Jupiter have short characteristic times and we may therefore conclude that the probability of *some* event involving Jupiter is very high and perhaps almost certain to occur. Yet we must not forget that orbital rounding which would be very rapid for Jupiter and Saturn. Thus many of the events of Table 10.2 will be impossible after about 100000 years since some orbital intersections will no longer exist. After three times this period the two most massive planets will have almost circular orbits and so we see that the most likely interactions are between A, B and Jupiter. The innermost planets would remain in very eccentric orbits for about six and two million years respectively and so we can safely assume that the probability of *some* major event between early planets is quite high.

10.5   A COLLISION

We have argued above that a natural consequence of the formation of the solar system with very eccentric orbits is the virtual certainty of events likely to cause major orbital changes following planetary condensation. In addition to the Jovian-type planets known today we would expect, by the Capture Theory model, the formation of two planets with orbits within that of Jupiter. The expected masses would be around that of Uranus for A and around five times that of the earth for B, the innermost. These values are speculative, but not unreasonable in view of the existing sequence of masses. Of course there is no direct evidence for planets A and B being present-day members of the solar system. Nor does there exist any known mechanism for their destruction other than as a result of an interaction. There is ample evidence for collisional events. Almost all the solid bodies which have been closely observed display impact craters which seem to have been caused by high-velocity collisions with solid objects. The view that the solar system has contained, from early times, a host of small bodies is inescapable. The asteroids, mainly between Mars and Jupiter, may be today's residue of a much larger population, it is possible that bodies lost in impacts are replaced by others from time to time. Further evidence of catastrophic events is provided by the smaller planetary satellites such as Phobos, Deimos, Amalthea, Hyperion, Miranda, and many others, which have irregular shapes suggestive of fragments of larger bodies. Indeed their appearance may be typical of many asteroids.

Now, the Capture Theory does not predict that direct formation of small solid bodies. The six, mainly gaseous, planets of Table 10.1 would be differentiated during the condensation process and so would possess solid cores. Consequently we would argue that a collision between two of them (A and B) is necessary to release the debris which has had such an important role in shaping the system. The high probability of such an event is clear from the last section. Apart from 'removing' the two planets involved, this collision, as we shall see below, enables a fair number of features of the solar system to be explained in a physically satisfactory manner. Indeed, when the full implications of a collision are reviewed, one is led to the conclusion that one *must* have occurred.

In 1977 we considered in some detail the consequences of a planetary collision between A and B in Table 10.1. It will be clear from the start that the event is somewhat different to any high-speed terrestrial impact. The effects of high-velocity impacts of small projectiles (e.g. bullets) with various targets is well-known and a number of calculations with velocities up to 6.25 km s^{-1} have been performed (Gault and Heitowit, 1963). It was found that about half the initial kinetic energy of a projectile was transferred to the material thrown out of the target (ejecta); the rest went mainly into heating and crushing the target. Extrapolating the results to a planetary collision is rather difficult for a number of reasons. Some of these are:

(a) very high masses give rise to much higher impact velocities than can be achieved in experiments, or free-fall on the earth or the moon;
(b) there is no clear distinction between projectile and target since the two planets would be of similar size;
(c) gravitational forces will be important in determining the trajectories of the collision products;
(d) both bodies will be rearranged and, to some extent, mixed thus implying gravitational energy changes.

Neglecting orbital velocities the minimum impact speed for our two planets is given by $W = \sqrt{[2G(m + M)/(r + R)]}$, where m and M are the two masses, with r and R the two radii of the planets. Inserting the values from Table 10.1 yields $W = 25$ km s^{-1}, a value four times as great as considered in calculations. Assuming that a fraction (around 30%) of the available kinetic energy were to go to heat material in the site of the impact, substantial vaporization is implied. In other words the impact would be explosive, the expansion of vaporized material injecting kineic energy back into the collision products.

A schematic representation of a head-on collision is shown in Fig. 10.8. Three stages are identified: stage (a) shows the situation soon after impact with high-velocity ejecta at small angles to the planetary surfaces. A shock wave travels out from the point of impact and stage (b) occurs when the smaller planet is completely shocked. At this stage a mass of vaporized material is trapped between the two planets. In stage (c) the expanding vapour has pushed apart the bodies with the smaller one being shattered. Subsequently the bodies would separate to take up distinct heliocentric orbits.

An actual collision would be oblique and so the cylindical symmetry of the head-on case would be lost. Fig. 10.9 illustrates an oblique collision in which the shear forces differentiate the lesser planet (stage (a)) forming two main streams following the break-up (stage (b)). Following this type of collision we would expect the massive ejecta to re-accumulate as a result of gravitational attractions. This type of behaviour has been noted in some recent SPH simulations by Benz et al. (1986, 1987). The outcome in this case would be three significant bodies: the damaged planet A and two major fragments of planet B eventually forming two new planets of terrestrial type. In addition there would be a very large number of much smaller solid fragments and much gaseous ejecta.

Although we have given a plausible picture of a planetary collision, a quantitative description is highly desirable. The SPH technique (Chapter 6) could be used in a computer simulation but this is very expensive, even when compared with our earlier encounter calculations. Yet it is fairly easy to gain some dynamical insights into the collision event by making use of the familar Newton's Law of Restitution, which

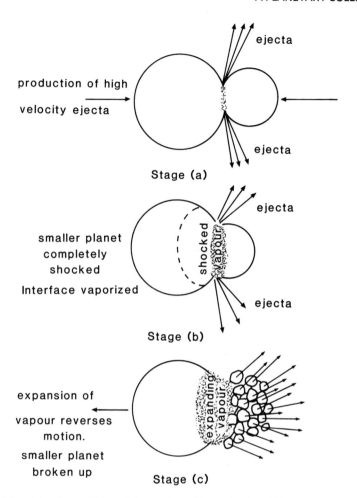

Fig. 10.8 — A head-on collision of planets, Stage (a): production of high velocity ejecta; stage (b): smaller planet completely shocked and interface vaporized; stage (c): expansion of vapour reverses motion and smaller planet is broken up.

might more commonly be applied to impacts involving billiard balls. The law states: when two bodies collide their relative parting velocity in the direction of the common normal at the point of impact is $-\varepsilon$ times their relative approach velocity in the same direction. The minus sign indicates the usual rebound following an imact. A one-dimensional case with elastic spheres is illustrated in Fig. 10.10 where three stages are identified. At the first stage the larger sphere is about to strike the small one ($U > u$). The second stage shows that instant of greatest compression, when some of the kinetic energy of the system has been converted to compressional energy. Some of this compressional energy will be restored to kinetic energy, pushing apart the two spheres. The final stage shows the bodies separating ($v > V$). The similarity to our above picture of the planetary collision of clear. Since the spheres are not perfectly elastic, some of the kinetic energy will be lost, according to the value of

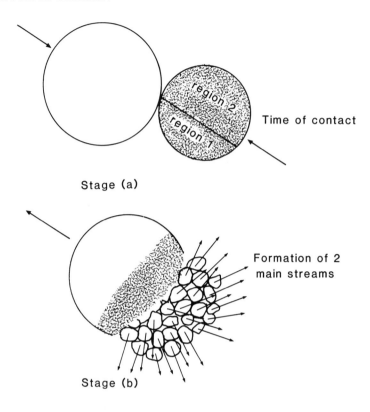

Stage (a)

Time of contact

Formation of 2
main streams

Stage (b)

Fig. 10.9 — An oblique collision between planets.

ε, the coefficient of restitution. The new velocities v and V are easily obtained in terms of u, U, and ε, since

$$v - V = -\varepsilon(u - U) \ ,$$

and by conserving of momentum,

$$mv + MV = mu + MU$$

Also the kinetic energy lost is found to be

$$\frac{(1 - \varepsilon^2)mM(U - u)^2}{2(m + M)}$$

In the case of billard balls, most of the lost kinetic energy would be converted to heat. For the planetary case much heating would occur also but some energy would be used

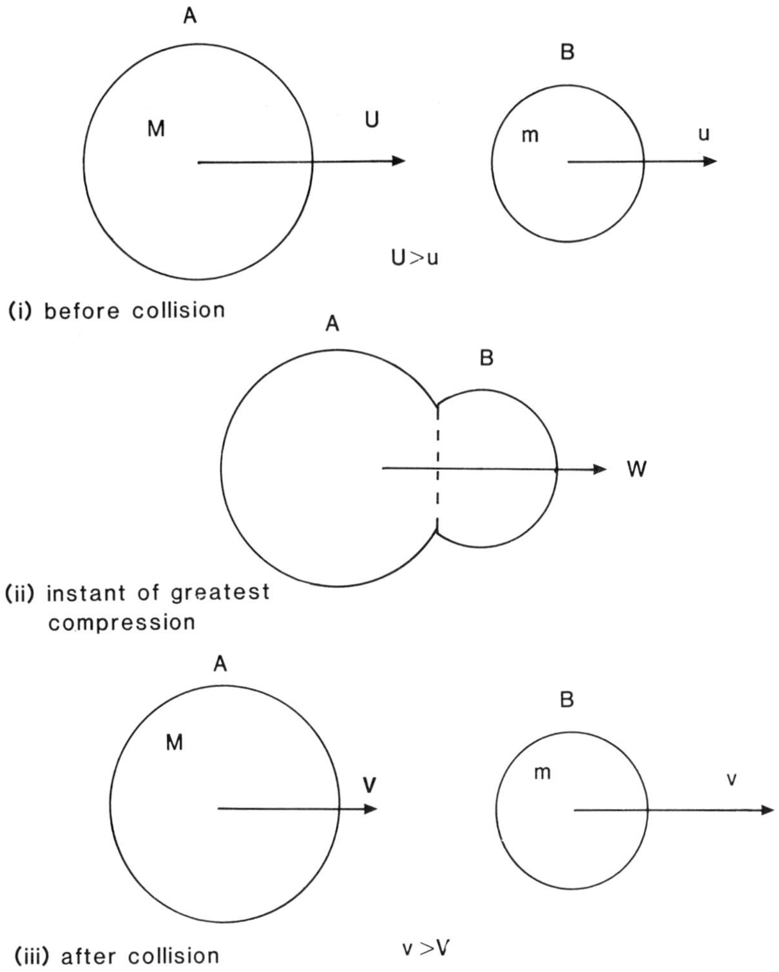

A

M   U

B

m   u

U > u

(i) before collision

A

B

W

(ii) instant of greatest
    compression

A

M
                               V

B

m   v

(iii) after collision   v > V

Fig. 10.10 — An elastic collision in one dimension.

in 'shattering' the smaller planet planet B. If we accept the computational result (albeit with very small masses and lower velocity) which suggests a loss of round 50% of the kinetic energy for the rebound motion, then it is reasonable to consider a value of $\varepsilon = 0.75$. We can then estimate the major components of the rebound velocities. What fraction of the energy would be used in the fragmentation process? This is uncertain, but in our 1977 paper we considered a case in which 70% of the non-rebound energy would provide separation velocities for the two major fragments of planet B. Fig. 10.11 shows the appropriate parameters of the collision; we assumed speed w normal to the impact direction for each of the fragments. The values of the various parameters are given in Table 10.3. In this example 87% of the original kinetic energy reappears as kinetic energy after the collision. The orbit elements of

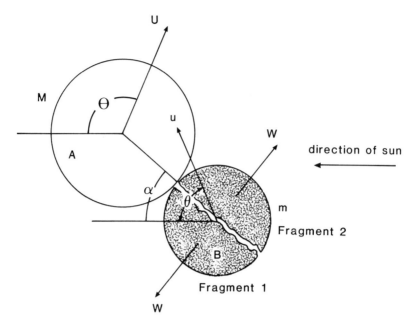

Fig. 10.11 — Collision parameters (see Table 10.3).

Table 10.3 — Pre- and post-collision orbits

Collision parameters (Fig. 10.11).
$M = 33.5M_\oplus$, $m = 5.3M_\oplus$, $U = 32.2\,\mathrm{km\,s^{-1}}$, $u = 35.8\,\mathrm{km\,s^{-1}}$, $\Theta = 100.1°$, $\theta = 49.7°$,
$\alpha = 60°$, $\varepsilon = 0.75$, Distance of sun = 1.58 AU.

Orbit elements Object	Semi-major axis a(AU)	Eccentricty e	Semi-latus rectum $p = a(1 - e^2)$
A before collision	17.53	0.91	2.93
B before collision	10.07	0.92	1.53
A after collision	—	1.002	—
Fragment 1	1.86	0.79	0.70
Fragment 2	1.80	0.64	1.06

the collision products may be calculated easily once the various velocity vectors are defined. However, the high masses involved imply perturbations large enough to invalidate two-body approximations and so the trajectories of the bodies were computed numerically until final orbit elements could be established. In fact the same technique was used to determine the initial orbits by computing backwards in time from the postulated impact.

10.6   RESULTS OF A COLLISION

The collision considered above is obviously just a single example which we present for the purposes of illustration; collisions with a wide variation of parameters are possible. Nevertheless we choose to discuss the results in Table 10.3 because they are particularly interesting with reference to actual solar system parameters. First we note that the larger planet (A) has achieved a hyperbolic orbit ($e > 1$) following the impact. This means that it leaves the solar system never to return. Second, we see that the orbits of the two fragments are in the terrestrial region. In this case the semi-latera recta are good approximations to those of Venus and the earth (the semi-latus rectum is the most persistent feature of an orbit being rounded-off in a resisting medium). These two fragments would arise from the solid core of the planet B and most of the gaseous component would be lost.

The event described here in some detail is not the only one which could give rise to terrestrial planets with appropriate characteristics. Variation of some of the parameters produces only moderate changes of outcome. It should be stressed that the present absence of planet A does not make its expulsion from the system an essential result of the collision. Since this planet crosses the orbit of Jupiter (whether perturbated greatly by the collision or not) it is subject to Jovian perturbation, perhaps for several million years. Our earlier calculation of characteristic times for major events suggests that a planet A would be a prime candidate for expulsion and it could have been expelled a considerable time after the collision event. However, in section 13.1 we suggest another possible outcome of the collision in which planet A, or at least its material, would have been retained in the system

We now turn our attention to the other major objects in the terrestrial region — Mars, Mercury and the moon. The similarity of the moon and the Galilean satellites of Jupiter is evident from Table 2.3 and so it seems not unreasonable to propose for all these bodies the same mode of origin. We would expect planet A, and perhaps also planet B, to possess some satellites (see Chapter 8) and we consider the moon to have been among their number. Is it possible for a satellite of A to be captured by one of the fragments of B (the earth) following the cataclysm? Fortunately this was an easy question to answer since we had already contructed a many-body computer model to investigate the motion of the collision products and so we needed only one extra body initially in orbit about A. It turned out that the orbital position of the moon relative to the line of impact is the crucial factor in determining a capture. If the moon had been on the appropriate portion of its orbit at the time of impact it would have been captured by the terrestrial fragment. The capture would be permanent, providing no further major interactions occurred. Had the moon not been in the 'right' place it may have been carried away by its parent planet or released in an independent heliocentric orbit.

Reference to Table 2.2 shows the planet Mars to be of similar density to the moon, although somewhat more massive, and so it is reasonable to consider it as another of the satellites of planet A which happened to become independent after the collision. Mercury is very different. Although its mass is reasonable for a major satellite it is much more dense than any known such body. In fact, when compression is taken into account, it is intrinsically more dense than the earth. For this reason we propose that Mercury is a high density fragment from the core of B.

The idea that asteroids were the products of the break-up of a planet is of long standing and has an obvious rationale. Observations of asteroids show that their brightness varies as they tumble through space, suggesting that they are of irregular shape — as one would expect if they were collision fragments. The usual image of an asteroid is that of an object similar in appearance to the Martian satellites, Phobos (Plate 2) and Deimos, which may indeed have asteroid origins. Another supporting piece of evidence is that it is generally accepted that meteorites are the result of inter-asteroid collisions, and their division into stones and irons (with very few of mixed type) ties in with ideas of the layered structure of a solid planet. Finally there is the gap in the solar system where Bode's law suggests that a planet ought to be found and it is in this region that the asteroids mainly exist.

Against this idea there has been expressed the view that there is no source of energy which could explain the spontaneous break-up of an isolated planet (see, for example, Napier and Dodd, 1973). Consequently there has been rather more support for the idea that the asteroids are the product of a 'spoiled planet'; one that did not form in the asteroid region because circumstances there were unfavourable. It has been suggested that tidal effects due to Jupiter might have prevented the accumulation of material to form a planet; according to the spoiled planet hypothesis there were formed in the asteroid-belt region numbers of small so-called 'parent bodies' of sub-lunar mass (Anders, 1971) and it is the debris from the break-up of these which give meteorites. By classification of meteorites in various ways, for example by the ratio of the abundances of various pairs of elements, meteoriticists try to estimate the probable number of original parent bodies.

A problem of the parent-body hypothesis is to find a source of heat which would explain the undoubted conclusion from observation that meteorite material had been molten at some stage — or even in a vaporized state. A solution, or at least a partial solution, to this problem was the discovery of the one-time existence of ^{26}Al, with a half-life of 700 000 years, which could have provided enough energy to melt even small bodies if there was enough of it around. Its detectable presence in meteorites is restricted to a few white high-temperature inclusions in carbonaceous chondrites, a type of stony meteorite, so its widespread occurrence in the early solar system must still be open to doubt.

The planetary-collision hypothesis does seem to answer all of the difficulties to which reference has been made. Since there are two bodies, the energy for break-up and dispersal comes from their orbital motions and gravitational interaction (in Chapter 13 we shall be considering other energy sources as well). The gravitational energy released by the collapse of the original quite-massive protoplanets would have been ample to have melted all of their substance, and their considerable gravitational fields would have led, on a short timescale, to the separation of material by density to give a layered structure. Masses of molten iron (and nickel) which were subsequently ejected by the collision and cooled to form iron asteroids would then give the characteristic Widmanstätten patterns which are observed.

There were two sources of molten silicate droplets which form chondrules in chondritic meteorites (section 2.5) and they were probably both effective. The first of these is the molten rocks which would have existed beneath the solid surfaces of the planets prior to the collision and the second would be droplets formed by condensation from a silicate vapour; we shall discuss later the possible thermal regimes which existed after the collision, but for now it is enough to say that if only a tiny fraction of the kinetic energy of the collision was turned into heat then there would have been an abundance of vaporized material. The total scale of the planetary collision in both space and time is modest compared with, for example, the cooling time of hypothetical solar nebula or the space and time required for a supernova-induced event. There is now strong evidence that the chondrules cooled quickly after their formation, which argues strongly that they were formed in a small-scale event, and also that chondritic meteorites were formed from material which had come from planet-sized bodies (Hutchison *et al.*, 1988). Chondrules would have become embedded in a matrix of small solid silicate fragments and subsequently consolidated in fairly large bodies (parent bodies) and thus chondritic meteorite material would have been released by subsequent collisions between parent bodies. The different clarity with which chondrules are seen, which leads to their petrological classification, would depend on the speed with which they solidified and the temperature of the surrounding matrix material. Thus if they were slowly cooling in a rather hot matrix one might expect some merging of boundary material and, consequently, a poorly defined chondrule.

10.7   CONCLUDING REMARKS

Following a planetary collision it appears that many of the 'irregular' features of the solar system can be explained in terms of simple and straightforward mechanisms. We include the terrestrial planets in the irregular category since they would not be likely to arise from the basic capture process which spawned the major planets. The Capture Theory is not unique in this dichotomy; many authors have experienced difficulty in specifying a process which could account for all the planets. Yet, unlike many other cosmogonies, the Capture Theory yields eccentric orbits which provide ample opportunity for second-order effects. In this chapter we have argued that the Capture Theory scenario makes a collision between two major planets in the early solar system almost inevitable. Such a cataclysm yields the broad spectrum of masses which is required by any reasonable consideration of this problem.

In its long history the solar system has been influenced by many events of an irreversible kind, and this makes it difficult to deduce what has been its state in previous times or what events actually took place. However, an event as cataclysmic as the one being postulated here, a collision between planets, might be expected to have left behind some indelible record and we shall now consider a number of features of the solar system, other than those already discussed, which can plausibly be linked to the collision event.

11

The moon and its characteristics

11.1 THE PHYSICAL STRUCTURE OF THE MOON

The obverse face of the moon is a well-known feature of the night sky and even with the most primitive telescope the main features of the lunar surface can be recognized (Plate 5). There are seen the great mare features long suspected, and now known, to have been caused by giant projectiles. There are the craters, caused by smaller impacts and without the infill of magma which gives to the maria their characteristic dark smooth surfaces. From some of the larger craters there emanate great light-coloured splashes of debris, called rays, and their presence indicates that the crater is of comparatively recent origin. Eventually, through the effect of bombardment by solar particles, and perhaps also thermal erosion caused by alternate solar heating and cooling, the rays become less visible and eventually disappear. Surrounding the maria, and indeed being the type of surface in which the maria were formed, are the lunar highlands, regions of lower-density anorthositic rocks. With a better observing instrument there can also be seen long crack-like features called rills. The largest of these are as long and deep as the Grand Canyon although, of course, they are not of fluvial origin. Such is the side of the moon that we see from earth.

The scenario we have suggested in Chapter 8 for producing satellites, in a filament left behind by a rotationally disrupted protoplanet, would give satellite collapse to high densities on a short timescale. This has implications for the initial thermal profile in satellites in general and in the moon in particular. As material rained onto the core of the forming satellite, so it would have struck the core with ever-increasing speed. If the material was of uniform density then, when the core reached a radius r, its mass would have been

$$M_r = \frac{4\pi r^3 \rho}{3} \tag{11.1}$$

and material would have fallen onto it with an intrinsic kinetic energy (kinetic energy per unit mass) of GM_r/r, which is proportional to r^2. If this energy was turned into heat at the point of impact, then the increase in temperature produced was approximately proportional to r^2. Such models have been considered by, amongst others, Toksöz and Solomon (1973) who also took into account radiative cooling during the formation process and the conversion of some of the projectile kinetic energy into shock waves which were transmitted into, and heated, the lunar interior. A typical thermal profile found for the non-gaseous part of a newly formed satellite is shown in Fig. 11.1, where the sharp fall-off in temperature at the surface represents

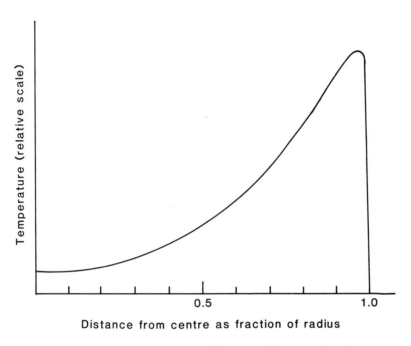

Fig. 11.1 — Initial thermal profile for a rapidly accreted satellite.

the effect of very rapid cooling by radiation. For satellites of radius greater than about 1100 km, a molten region would have formed close to the surface, and this would certainly have been true for the moon. We have carried out calculations of the thermal evolution of the moon, including the effects of conduction, convection, radiation and heat production by radioactivity. The results of one such calculation are shown in Fig. 11.2. It properly predicts the existence of a partially melted region at the centre of the moon at present, but also shows the way in which the molten zone, which began near the surface, gradually retreated into the interior.

The implications of this model for maria formation are fairly obvious. A projectile falling on the lunar surface with sufficient speed would produce a large basin and also create cracks reaching down to the interior regions. It is through these cracks that the subsequent volcanism would have taken place. Later we shall refer to this again in somewhat greater detail.

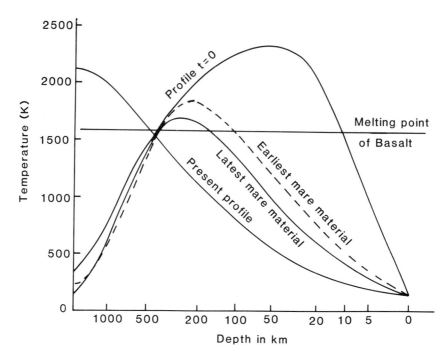

Fig. 11.2 — The thermal evolution of the moon, including the effects of radioactive heating.

In 1959 the Soviet Union sent on a journey around the moon its Lunik spacecraft, which transmitted to the earth the first pictures of the reverse side. The pictures were of poor quality, but they were good enough to show, rather surprisingly, that the appearance is quite different from that of the side we normally see. The whole hemisphere is dominated by highland regions and, while mare features are not completely absent, those that are present are of small extent (Plate 6). This immediately posed the question of *why* the two sides are different, and the first thoughts were that for some reason the farside of the moon had been shielded from the impacts of large projectiles. Altimeter readings by later spacecraft soon negated that explanation; in fact there *have* been large impacts which have caused extensive basins, but they have never been filled by molten magma from below. Eventually came the answer, which seems a very reasonable one and is supported by some seismic data, that the crust of the moon is thicker on the farside than on the nearside, 100 km against 50 km, so that cracking of the solid surface material did not extend into the molten regions of the moon. However, that raises the new question of why the difference of crustal thickness should occur and some explanations which have been given, based on dynamical arguments, are far from convincing.

A satellite in the vicinity of the postulated planetary collision would have been directly affected by it in a relatively short time. Theoretical and experimental work on hypervelocity impact shows that the ejecta from such an event travels with respect to the centre of mass of the system at some three times the approach speed of the two bodies (Gault and Heitowit, 1963). From their gravitational attraction alone, even discounting the approach speeds due to their orbits, the planets would have collided at a speed in the region of 50 km s^{-1} so that debris at speeds up to 150 km s^{-1} would have resulted. For any reasonable satellite distance (around 500 000 km) the debris would have arrived within an hour or so and we now discuss what effect it would have had on the satellite. First we consider the onfall of a slow projectile on a solid astronomical body; that is to say, a projectile with a speed little more than the escape speed from the body. The effect is illustrated in Fig. 11.3(a); the projectile shares its

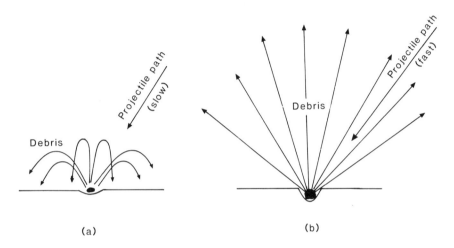

(a)   (b)

Fig. 11.3 — The effect of a projectile falling on a solid-body surface. (a) A slow collision gives accretion of the projectile. (b) A fast collision gives abrasion of the body.

energy with surface material, some is converted into heat but the debris all moves at less than escape speed, falls back on to the surface and the projectile is thus *accreted*. Now, at the other extreme, we consider the onfall of a projectile with a speed many times that of escape speed from the body. This is illustrated in Fig. 11.3(b); even allowing for conversion of some energy into heat, much of the debris has greater than the escape velocity and so leaves the body. The mass leaving can be many times greater than that of the projectile so the net effect is that the body is *abraded*. The escape speed from the moon is 2.4 km s^{-1} so the projectiles we are considering would have had speeds 60 times the escape speed and, in principle, be capable of removing up to 3600 times their own mass of lunar material. We are suggesting that one hemisphere of the moon, that which faced the original parent planet and for the same dynamical reason now faces the earth, was abraded by the ejecta from the collision, thus giving the observed difference of crustal thickness of the two sides.

It is possible to make an order-of-magnitude estimate of the mass of ejecta necessary to do this. The total mass of one lunar hemisphere of crustal material of thickness 50 km is approximately 2.5×10^{21} kg. Assuming that the abrasion efficiency was 10%, so that a unit mass of ejecta would abrade 360 times its own mass, then the mass of ejecta required to fall on the moon would be around 7×10^{18} kg. If the original orbital radius of the moon was, say, 500 000 km, then it would subtend a solid angle 1.2×10^{-5} steradians at the collision. For an isotropic distribution of ejecta from the collision (very improbable) this would require a total loss of ejecta material of 7×10^{24} kg, or slightly more than the mass of the earth. The calculation is a rough one, but good enough to show that the abrasion hypothesis does not require an impossible mass of planetary debris. A more likely value for the required mass of ejecta, taking into account anistropy of its directions of motion, is much less than the mass of the earth.

Larger debris from the planetary collision, released into heliocentric orbits, would, over a long period of time, have given rise to the observed cratering of bodies in the solar system; presumably this would have been at a decreasing rate as the supply of such debris was exhausted by the collisions themselves. The evidence from the moon is that the main period of bombardment and volcanism covered a period of about 700 million years, with the earliest magmas dating from 3900 million years ago. The volcanism might be older than that since earlier volcanic effluents would have been covered over by later events.

We now consider the sequence of events producing a typical mare. A large projectile striking the moon would have gouged out a large and deep basin, with a profile as illustrated in Fig. 11.4(a), and also caused rifting and cracking of material

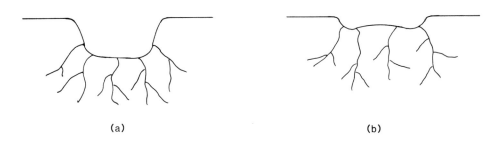

(a)   (b)

Fig. 11.4 — The evolution of a lunar basin. (a) The initial projectile fall produces a large deep basin and cracks the lunar crust. (b) Internal pressure forces up the floor of the basin.

to a considerable depth. However, this configuration would have been unstable. The hydrostatic pressures on the floor of the basin would have created compressive stresses in the material too great for it to withstand, and the floor would have adjusted itself upwards until the depth was 10–20 km, at which point the material would have been strong enough to resist the forces on it. This state of affairs is shown in Fig. 11.4(b).

What can be deduced from the cooling curves given in Fig. 11.2 is that the main cooling was taking place in the outer material while the interior, subjected to radioactive heating, became hotter. Consequently the outer solid region became in a state of tensile stress due to the cooling — it would have liked to shrink — and at the same time it subjected the molten material in the interior to an extra component of pressure which was not of hydrostatic origin. Periodically the stress and pressure were relieved simultaneously by volcanic eruptions as the solid crustal material yielded and molten material under suprahydrostatic pressure was ejected. The regions where the yielding took place were within the mare basins where the solid crust was thinner and where the projectile had also formed cracks in underlying layers.

At this stage it is necessary to describe isostacy, which is one of the fundamental concepts of geophysics. The basis of this idea is that there is a uniform mass per unit area over the surface of the earth above some level, called the compensation level, in the interior. There is more than one model for the way that isostatic equilibrium is attained, but for our purposes it is enough to know that the principle is expected to operate for any large solid body — such as the moon. Thus, since the mare lavas are denser than the surrounding highland anorthositic rocks, it was expected that the lower level of the mare surfaces would be governed by isostacy. It was a surprise when an analysis of the motions of lunar orbiters showed gravitational anomalies in the mare regions which indicated an excess of mass over and above what was required for isostacy; these excess concentrations of mass have been dubbed *mascons*. The first thoughts were that they represented the buried remains of iron projectiles that had produced the basins, but it was soon realized that this could not be so. Since there had to be molten regions not too far below the surface when the projectiles fell, the strength of the solid layers would not have supported the extra weight and the projectiles would have sunk towards the centre of the moon. In any case, analysis of the excess gravitational field has revealed that the extra mass is probably in a disk-like form and so is almost certainly an overfill of the mare basins to above the level required for isostacy.

The scenario we have presented for mare formation is consistent with the mascon observations. Molten matter would have been expelled under the excess suprahydrostatic pressure due to the cooling and shrinking surface regions. This matter would quickly have solidified and contributed a local hydrostatic pressure greater than that corresponding to isostacy. The surface would then have slowly slumped to reduce this excess pressure, resisted only by the strength of the solid crustal material, which would have been capable of supporting *some* excess pressure. The next bout of volcanism would have repeated the process, but the cycles of eruption and slumping would have become more and more sluggish as the volcanism reduced in scale and the solid region became thicker and mechanically stronger. The end of the process would be mass excesses similar to those observed.

This completes our survey of the major features of the moon and how they might be related to the planetary collision. This scenario has been subjected to computational modelling which yields results supporting the effects which have been described here.

11.2   THE MOON AND MAGNETISM

The basalts which flooded the great lunar basins between about 3.9 and 3.2×10^9 years ago contain small particles of free iron, and for this reason they preserve a magnetic record of the time when they solidified. At a temperature above what is

called the Curie point, ferromagnetic materials do not become permanently magnetized by the effect of an external magnetic field. At a lower temperature they become
magnetized to an extent which increases, but not linearly, with the external field and
then if the external field is removed, they remain magnetized and are, in effect,
permanent magnets. The Curie temperature for pure iron is 1043 K and, once the
solid basalt had cooled and remained below that temperature, the strength of its
magnetization, the so-called natural remanent magnetization (NRM), tells us the
strength of the surface magnetic field of the moon when it solidified. Analysis of lunar
samples shows that some of the oldest basalts cooled in a surface field of 10^{-4} T
which steadily declined over the period of lunar volcanism (Stephenson *et al.*, 1975).
The variation of the lunar magnetic palaeofield with time is illustrated in Fig. 11.5.

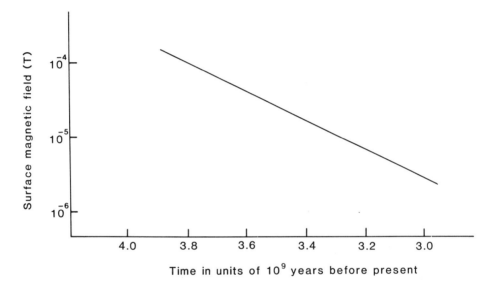

Fig. 11.5 — The change of the lunar magnetic palaeofield with time, (after Stephenson *et al.*,
1975).

The significance of the lunar measurements can be judged by comparison with
the earth's magnetic field which, over the surface, varies between about 3 and
7×10^{-5} T.

The earth's field is very approximately like a dipole field which would result from
a magnet situated at the centre of the earth (Fig. 11.6). However, despite the
intensity of magnetization of its surface rocks, the moon has an almost undetectable
field as measured from fairly distant spacecraft, and near the surface local fields of
only a few hundred nT (1 nT $= 10^{-9}$ T) are detected, which are variable in strength
and direction on a kilometre scale.

The accepted theory for the magnetization of the earth, and other planets, is that
it is due to a dynamo mechanism operating in molten conducting cores. This has led
Runcorn (1975) to propose that the early lunar magnetic field was produced by a
dynamo mechanism in a molten core of radius about 500 km but that as the core

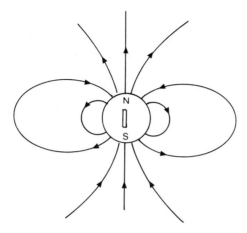

Fig. 11.6 — An approximate representation of the earth's magnetic field.

solidified so the field diminished, as is seen in Fig. 11.5. The pattern of NRM in the moon from such a scenario would lead to a zero external field, which is observed; on the other hand, if the moon had been magnetized by a steady external field then the NRM pattern *would* give an external field. The patterns of surface NRM from the two situations are shown in Fig. 11.7.

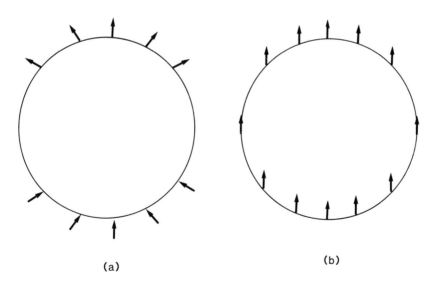

(a)   (b)

Fig. 11.7 — The form of surface NRM expected for: (a) magnetization by an internally-generated dipole field, (b) magnetization by an external uniform field.

There are some problems with the Runcorn theory. It has been shown by Stock and Woolfson (1983a) from the known moment-of-inertia factor of the moon that it is impossible for it to have an iron core greater in radius than about 385 km, which is rather small for the fields which need to be generated to explain the observations. It *is* possible that the core could be, in whole or in part, other than pure iron — for example, it could contain iron sulphide (troilite) — but that would seem to be rather an *ad hoc* solution to the problem as there is little evidence for the existence of troilite on such a scale in the solar system. Another difficulty is to find a source of energy which can give an active molten core for the period required. Here Runcorn has suggested heating by radioactive 'superheavy' elements in the early solar system, although attempts to find evidence of the existence of such elements have actually led to the conclusion that they did *not* exist. This conclusion is based on the examination of unmetamorphosed old meteorites with metal–silicate interfaces. If there had been superheavy elements in the iron then particle tracks would have been observed in the silicate, and since it has never been reheated there is no way that such tracks could be removed by an annealing process. The lack of such tracks is evidence of the absence, or at least an undetectably low level, of superheavy elements in the samples examined.

We now turn to another possible explanation offered by Stock and Woolfson (1983b). The model of satellite formation as a fast collapse in a filament condensation leads naturally to the initial thermal profile shown in Fig. 11.2, which also shows calculated thermal profiles at the beginning and end of the period of lunar volcanism. It will be seen that the molten region is 100 km below the surface at the beginning of the period, retreating to 150 km by the end. Changing the parameters of the computational model can change these depths, but the values we have here will do for illustrative purposes. The effect of capture by the earth into a highly eccentric orbit would have had a profound effect on the lunar interior. We know, for example, that the very small deviation of Io, the innermost Galilean satellite, from a circular orbit due to perturbations by Europa, the next one out, causes sufficient energy generation within Io to give volcanic activity. The small changes of distance to Jupiter during an orbit gives varying tidal stress so that Io is constantly being stretched and then relaxed. Since it is not a perfectly elastic body there are hysteresis losses so that in each cycle some of the energy of stretching is converted into heat. The early earth–moon system would have behaved similarly; although the earth has only 1/318 of the mass of Jupiter, the initial eccentricity of the lunar orbit would have been quite high so that heating effects comparable to, or even greater than, those now observed for Io might have been present.

A characteristic of tidal effects is that they are at a maximum in the plane of the orbit of the distorted body; thus the effects would have been greatest in the region of the lunar equator assuming that, as now, the spin axis of the moon was perpendicular to its orbital plane. While there would have been a molten zone below the surface everywhere, in regions below the equator the zone would have been at a higher temperature and closer to the surface. Actually there is some evidence for this. In Table 11.1, in the first two columns, there are shown the latitudes and longitudes of the centres of the largest mare basins; the positions of some of them can also be seen in Plate 5. It will be noticed that these centres show a strong tendency to lie within 30° of the equator, the exceptions being the centres for the largest one of all, Imbrium, and the large southern mare Australe. However, it turns out that the source region for the Imbrium flows can be identified as near the crater Euler and this is at a latitude of 23°. Despite the fact that the deepest penetration of the crust must have been in the

Table 11.1 — Positions of the centres of the major mare basins or the likely source positions for volcanism. In the final column, revised latitudes are given with respect to an equator tilted at 6° to the present one

Mare	Latitude (°)	Longitude (°)	Revised latitiude (°)
Imbrium (centre)	40 N	22 W	38 N
Imbrium (source)	23 N	29 W	20 N
Serenitatis	26 N	18 E	28 N
Tranqillitatis	10 N	28 E	13 N
Crisium	17 N	58 E	22 N
Marginus	18 N	82 E	24 N
Smythii	1 N	88 E	7 N
Foecunditatis	3 S	48 E	1 N
Nectaris	15 S	33 E	12 S
Nubium	19 S	16 W	21 S
Humorum	20 S	42 W	24 S
Procellarum	5 S	48 W	9 S
Orientalis	19 S	170 W	20 S
Moscoviense	25 N	145 E	28 N
Australe (centre)	45 S	90 E	39 S
Australe (likely sources)	35–42 S	90 E	29–36 S

centre of Imbrium, the magma came from near its southern edge — which is exactly what would have happened if there were higher temperatures (and lower viscosities) and less-deep material in the equatorial regions.

There is a similar situation for Australe. The basin centre is well to the south, but the main flows are concentrated in the north of the basin. The figures given for the range of possible source latitudes spans the bulk of the basalt region, just ignoring narrow fingers of flow away from this. In relation to a postulated original equator tilted at 6° to the present one, all the sources of volcanism could have been within 30° of the equator.

We now consider the earth–moon system, newly created in the planetary-collision event, moving around the sun in a highly eccentric orbit which would round off in a few tens of millions of years. It would have been influenced by the magnetic field of the very young sun, formed a few million years before the capture event. There has been a great deal of theoretical consideration of the expected field of the early sun. It is generally agreed that it would have been much higher than the present field — for example, about 10^3 times its present value with a decay time between 10^8–10^9 years is well within the range of speculation. Freeman (1978), considering the likely value of the primordial solar magnetic field from several points of view, comes up with a range of values between 12 and 6000 times the present value.

The pattern of the solar magnetic field is very different from that of a bar magnet, a dipole field, such as is shown in Fig. 11.6. Because it is a hot and active body, the sun emits a constant stream of charged particles, the solar wind, and the motions of these are influenced by the solar field and, in their turn, they influence the form of the solar field. While a dipole field varies in strength with distance, r, as r^{-3}, the distorted solar field varies at distances of more than a few solar radii as r^{-2}, with the field lines stretched out so that they tend to lie closer to the equatorial plane. Another, and important, feature of the solar field is its sector structure, illustrated in Fig. 11.8;

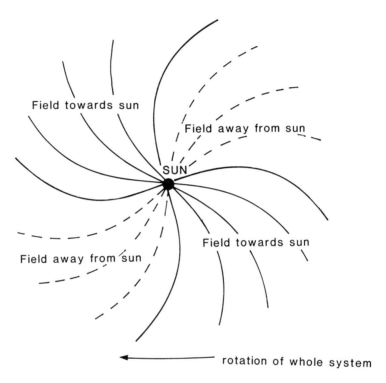

Fig. 11.8 — The sector structure of the solar magnetic field. The field lines alternately point away from and towards the sun in neighbouring sectors and the field is strongest in the equatorial plane.

looking down the rotation axis of the sun the pattern is that there are alternating regions where the magnetic field points approximately towards the sun and away from the sun.

It is now necessary to describe the relationships between electric currents and magnetic fields. The first of these is that if there is a change in the magnetic field traversing a ring conductor then an electric current is induced in the ring. The second is that a current flowing in a conductor creates a magnetic field in its environment. We now relate this to the passage of the earth–moon system around the early sun. The hot material below the lunar surface was predominantly silicate rocks and we are accustomed to think of such material as an insulator, or at least as a poor conductor of electricity. In fact it is what is known as a semiconducting material and it has the characteristic that its electrical conductivity increases with increasing temperature. When molten, most silicate rocks conduct electricity extremely well (Fig. 11.9). For

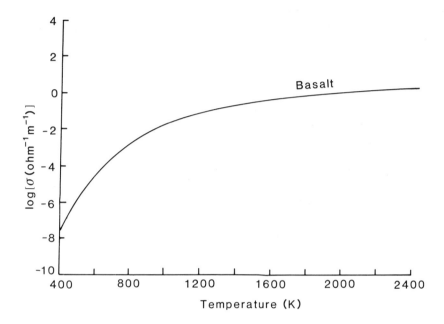

Fig. 11.9 — The variation of electrical conductivity with temperature of basalt, a semiconducting material.

that reason, with its highest subsurface temperatures below the equator, the early moon would have been a good approximation to a ring conductor moving in the solar field. That being so, due to the variations in magnetic field from place to place, currents would have been induced below the lunar surface and most strongly below the equatorial regions. Such currents would then have induced magnetic fields in their vicinity and surface material would have been magnetized by the combination of the solar field and the induced field.

There would have been two main mechanisms for the variation of the magnetic field in the vicinity of the moon — firstly the elliptical orbit of the earth–moon system around the sun, and secondly the sector structure of the solar field which rotates with the sun. It has been shown by Stock and Woolfson (1983b) that the induction processes described above could have given enhancement of the surface magnetic field by a factor up to five times that of the sun alone. With the solar field 1000 times its present strength, which is within the range of previous estimates, and an earth–moon heliocentric orbit of eccentricity 0.8 and semi-major axis 2.7×10^8 km, surface fields of well over 10^{-4} T would have been produced. The actual strength of the surface field, according to this model, varied with time and with position on the lunar surface, being strongest near the equator.

The major features of the magnetic observations are explained by this model. From the time of its eruption, magma would have taken of the order of 20–30 days to cool to a temperature below the Curie point; this being assisted by the fact that molten lunar material has a very low viscosity and would quickly spread in the form of a thin sheet. A factor which could have associated episodes of volcanism with the very highest fields is that eruptions might have been triggered, other things being favourable, at the time of a perihelion passage when there was the greatest rate of change of tidal stress. In support of this, it is observed today that moonquakes, feeble as they are, are most common at the time of perigee and apogee, when there is the greatest rate of change of tidal stress due to the earth. Another feature is that the direction of magnetization would be constantly changing due to the sector structure of the solar magnetic field. During a perihelion passage, with the sun rotating even at its present rate, the moon could easily have passed through of the order of ten sectors while an active eruption event was in progress. In this way the pattern would emerge of strongly magnetized surface material with variation of the surface magnetic field on a kilometre scale, dependent on the flow of magma and the rate at which it cooled through the Curie temperature, and an absence of any net field at larger distances from the moon.

This scenario has been explained in relation to the origin of the earth–moon system as a product of the planetary collision. However, if the system had originated in some other way then the mechanism described for producing surface magnetism could still apply.

12

Features of Mars and the outer planets

12.1 MARS — ITS ORIGIN AND EVOLUTION

We have seen that planet A, had it survived, would eventually have rounded off in the asteroid-belt region and its expulsion from the inner solar system explains the gap at 2.8 AU. Planet B would have occupied the Mars slot, except that its material was consumed in contributing to the earth, Venus and much other 'debris'. How then can we account for Mars, a rather small body but one which approximately fills a slot in the Bode's law sequence?

In Fig. 12.1 we show the densities of some of the smaller solid bodies in the solar

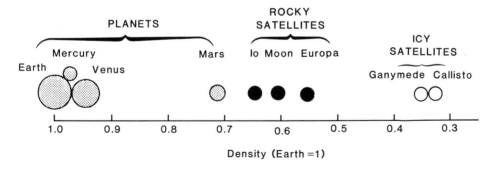

Fig. 12.1 — The densities of some small solid solar-system bodies.

system. We advance the idea that Mars has its origin as a satellite of one of the colliding planets; its density is suggestive of such an origin, although its mass is about four times that of Ganymede or Titan. However, the mechanism described in Chapter 8 for the formation of regular satellites would lead to large, and perhaps numerous, satellites for the innermost planets. It is these planets which would have been subjected to the largest solar-tidal forces and have acquired the greatest specific angular momentum in their tidal-bulge regions (section 8.5). Computer modelling shows that possible outcomes for satellites of the inner planets after the collision include escape into an independent heliocentric orbit, which is what we are suggesting for Mars.

A feature of Mars which supports this hypothesis is that its surface, like that of the moon, shows hemispherical asymmetry (Fig. 12.2.) The two regions differ in height

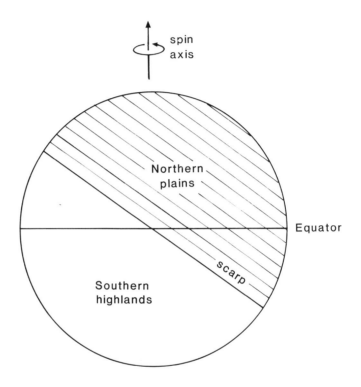

Fig. 12.2 — A schematic representation of the hemispherical asymmetry of Mars.

by about 4 km and are separated by a scarp, 2 km high, which runs round the body at 35° to the equator. The northern region, at the lower level, is smooth and lightly cratered, much like lunar maria, while the southern region is heavily cratered and resembles the lunar highlands. This pattern has been explained by Wise *et al.* (1979) as due to internal convection effects which have removed the crust in the northern region by mantle overturn. The basaltic material replacing the crust, being of higher density, would then be at a lower level in order to give isostacy.

Another feature of Mars which it shares with the moon is an offset between the centre of mass and the centre of figure (CM–CF offset). The centre of figure is where the centre of mass would be if the whole of the body was at the same density; thus the offset says something about density distributions within the body. In the case of the moon, the CM–CF offset is about 2 km and is completely explained by the removal of crust material and the readjustment of interior material to give a configuration of minimum energy. In the case of Mars, the CM–CF offset is about 2.5 km (Arvidson *et al.*, 1980). It is approximately normal to the plane of asymmetry, which suggests a linkage between the surface features and the offset.

The formation of the Martian surface features and the CM–CF offset have been explained in terms of the collision hypothesis by Connell and Woolfson (1983). The original northern hemisphere, which faced the colliding planets, was modelled as having layers of ice, silicate and mantle (see Fig. 12.3); due to the collison the ice and

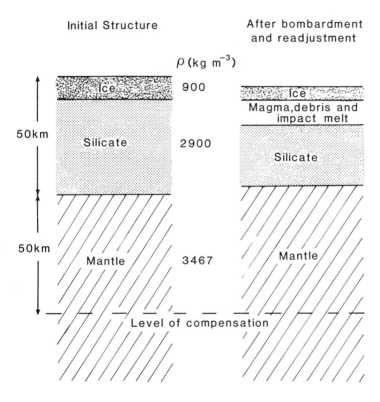

Fig. 12.3 — The postulated structure of Martian near-surface material before and after the planetary collision. The total pressure at the level of compensation remains unchanged.

some of the silicate was removed by abrasion and eventually replaced by denser basaltic material. It should be noted that any model of an early Mars must include considerable quantities of water, for which there is abundant evidence in surface features that are clearly formed by flowing water. Estimates for the amount of water, either in free form or as hydrated silicates, varies from an equivalence of 10 m to 1 km over the whole planet (Owen and Bieman, 1976; McElroy *et al.*, 1977; Anders and Owen, 1977; Allen, 1979). Based on the idea of Mars as a one-time icy satellite, resembling Callisto or Ganymede, Connell and Woolfson took a range from 160 m to

10 km in their models, although some of the ice could have been incorporated in the silicate layer.

Given the constraints of the densities of materials, the actual depression of the northern plain and the need for isostatic equilibrium there were a range of possible crustal modifications. For example, with a 1 km-thick ice layer (giving 49 km of silicate — see Fig. 12.3) it was necessary to remove 21.6 km of silicate and 60% of the original ice content and then to add 22.3 km of basaltic material. It turns out that this general pattern prevails for a very wide range of assumed initial conditions. Another result, which was somewhat unexpected, is that the calculated CM–CF offset was very insensitive to the initial assumptions and was between 2.81–2.86 km, in fairly good agreement with the quoted observationally deduced value of 2.5 km.

Because of its size, the volcanism on Mars, which added the extra surface material, would have lasted to a period when the frequency of projectile impacts had much reduced, thus explaining the paucity of craters in the northern region. The volcanism would have generated a great deal of heat and so liquefied and vaporized water to create a hydroclimate. One mystery about Mars is that, while it clearly had a great deal of water associated with it at some stage, the water has now mostly disappeared. With its present gravitational field and temperature, Mars should retain water indefinitely and this is certainly true if, as seems likely, the polar caps are completely or mostly water-ice. The dominant factor which governs the retention of an atmosphere is the value of

$$\psi = v_e^2/v_\theta^2 \qquad\qquad\qquad (12.1)$$

where v_θ is the root-mean-square velocity of a molecule and v_e is the escape velocity from the planet. If ψ is as low as 20 then the atmosphere will be lost fairly quickly; if it is as high as 60 then the atmosphere will be retained almost indefinitely. Intermediate values will give atmospheres that will be lost on timescales which could be as long as hundreds of millions of years or even more. The present ψ for water vapour on Mars is about 72, which retains water indefinitely, but if the temperature had ever been as high as 360 K for any long period of time then ψ would be reduced to 50 and water vapour would gradually have been lost. Such a high temperature would have enabled water to be in liquid form and so give the fluvial features observed on the Martian surface while at the same time giving a steady loss of atmospheric water. The idea has been advanced that Mars once had a higher temperature than now due to a CO_2-induced greenhouse effect (Pollack, et al., 1987) with the temperature gradually falling as the CO_2 was washed out of the atmosphere by water precipitation and incorporated as carbonates into solid surface material.

A feature of the moon is that its spin axis lies in the plane of asymmetry, which is not the case for Mars. The reason for this is that the moon became tidally coupled to the earth while Mars became a comparatively isolated body. The moon is slightly pear-shaped with the 'sharp' end pointing towards the earth as it would have done for its original parent. Even if the moon had been captured in a different configuration, it would have evolved by dynamical forces to its present state. However, if Mars was also a satellite then its spin axis would have been originally contained within the plane of asymmetry no matter what was its orientation in space when it became an isolated body. The spin axis is actually at an angle of 55° to the plane of asymmetry and Connell and Woolfson explained why this is so in terms of the concept of polar wander, which causes the surface to move relative to the spin axis. Polar wander has happened on the earth — it is an aspect of plate tectonics — and Runcorn (1980) has suggested it for the moon to explain some magnetic observations. Suggestions of polar wander on Mars have also been made from time to time to explain various surface observations (see, for example, Murray and Mallin, 1973; McAdoo and Burns, 1975).

If Mars, like the moon, had molten material close to its surface early in its history, then the solid lithosphere would have rested on a base of low viscosity material. Lamy and Burns (1972) have given a theorem which states that a spinning body, with internal dissipation of energy, will eventually move its spin axis until it corresponds with the axis of maximum moment of inertia. The principle of the theorem is very simple. For a body with moment of inertia I and angular momentum H the energy of spin motion is $H^2/(2I)$. If the body is losing energy but retaining angular momentum, then I has to increase, which is tantamount to saying that spin axis must move so as to get mass as far as possible from itself. Connell and Woolfson modelled the major surface features on Mars; some like the northern plains and the Hellas basin being depressions, while others like the Tharsis uplift and Olympus Mons were elevations. From the model it was calculated that the spin axis was at 11.9° from the axis of maximum moment of inertia of the modelled features. To get that close, or a closer, coincidence of the two axes by chance is not very likely (probability about 0.02), which supports the idea that the solid crust had moved over the remainder of the body to satisfy the requirements of the Lamy and Burns theorem. There are some uncertainties in the model used — for example, the Argyre basin might be a late feature so that the lithosphere was immobile when it was formed. Again, it is possible that the mobility of the lithosphere may have not persevered long enough for complete adjustment of the surface to have taken place. Finally it should be noted that Martian plate tectonics has been invoked to explain features such as the Valles Marineris, a great fissure in the Martian surface, which could be the location of an interplate boundary (Guest *et al.*, 1979). The form of polar wander suggested by Connell and Woolfson could have been produced by such a mechanism.

12.2   URANUS, NEPTUNE, TRITON AND PLUTO

The outermost reaches of the solar system contain the planets Uranus and Neptune, bodies of similar mass and external appearance whose characteristics suggest substantial rocky-metal cores surrounded by hydrogen and helium together with volatile compounds of hydrogen, oxygen, nitrogen and carbon.

The most curious feature of Uranus is its spin axis, which is inclined at 98° to the normal of its orbital plane (see Table 2.3). This, and the inclinations of the spin axes of the other major planets, are illustrated in Fig. 12.4. The satellite family of Uranus is regular in the sense that the orbits are in the equatorial plane of the planet, but the total arrangement is hardly one that could have come about solely through tidal interaction with the sun in the way proposed in Chapter 8. However, an explanation is readily available in terms of the early motions of the protoplanets. The analysis given in Chapter 10, which led to the realization that interactions were possible in the

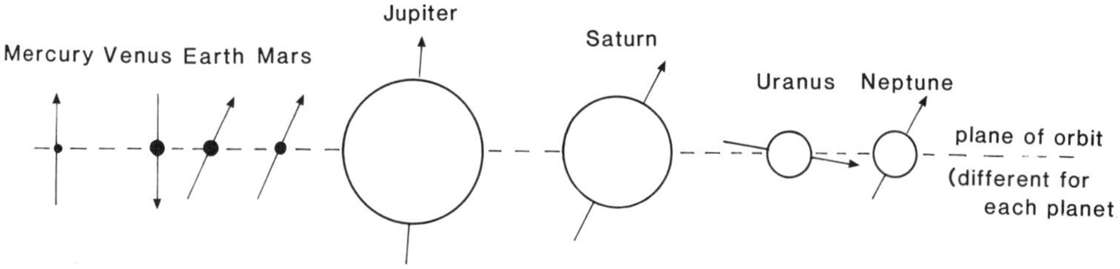

Fig. 12.4 — The relationship of spin axes to orbital planes for the planets.

early solar system, was mainly concerned with strong interactions which would lead to a catastrophic event such as the break-up of planets or a planet being ejected from the solar system. What is clear is that less severe events were also possible and would have had even higher probabilities than the catastrophic ones. Protoplanets passing by each other could have given tidal interactions of comparable strength, or even stronger, than those due to the sun. Since tidal effects are directly proportional to the tide-raising mass and inversely proportional to the *cube* of its distance, it turns out that Saturn at a distance of 0.7 AU has a stronger tidal effect than the sun at 11 AU, the perihelion distance of the initial Uranian orbit according to the model in Chapter 8. An early interaction, with closest distance, say, 0.5 AU, between Uranus and Saturn, with the vector of closest passage approximately normal to the orbital planes could have given the axial tilts of 98° and 27° for Uranus and Saturn respectively. There are other possible interactions to explain the observations; the point being made here is that the model of the early solar system with protoplanets in elliptical near-coplanar orbits gives the essential ingredients for an explanation, albeit not a unique one.

The spin axis of Neptune, tilted at about 29°, may also be ascribed to some distant interplanetary interaction, but the main peculiarity of this planet lies with its companions, the satellites Triton and Nereid. The former satellite is a large and massive one and would be difficult to explain in terms of the model in Chapter 8 since the initial perihelion distance of Neptune (around 17 AU) would not permit sufficient angular momentum to be transferred to its tidal bulge to give a large satellite. However, the most curious feature of Triton is that its orbit is *retrograde* and therefore it cannot be considered as a natural satellite in any case. The other satellite, Nereid, is fairly small (diameter about 300 km) and in a prograde orbit, but one of exceptionally large eccentricity (0.75).

In Chapter 2, when Pluto and its satellite Charon were described, it was suggested that Pluto could not be regarded as a normal member of the family of planets. Apart from its small mass, one-fifth that of the moon, its orbit is of high eccentricity (0.25) and high inclination (17°). In addition its perihelion distance, 29.63 AU, is just within that of Neptune, 29.81 AU, although because of the differing orbital inclinations and the commensurability of the orbital periods (ratio approximately 3:2) they never approach closely. The relationship between the orbits of Neptune and Pluto and the nature of the orbit of Triton have prompted speculation about relationships between these three bodies. Harrington and Van Flandern (1979) have presented a scenario in which Triton and Pluto were both satellites of Neptune and were disturbed by an unidentified planet of mass $5M_\oplus$ which passed through the system. The gravitational effect could have reversed the orbit of Triton, expelled Pluto into an independent heliocentric orbit, and tidally disrupted Pluto to give its satellite companion. This is dynamically quite feasible, although the authors do not justify the existence of the passing planet except for the purpose of disturbing the Neptunian system.

We have put forward an alternative scenario which is completely consistent with the Capture Theory model and planetary collision hypothesis (Dormand and Woolfson, 1980). We start with Neptune having two satellites, Pluto and Nereid, both small and in prograde orbits as is consistent with the moderate solar tidal effect on the early Neptune. Triton is now cast in the role of one of the large satellites of either planet A or planet B. Following the planetary collision it is released into a heliocentric orbit of high eccentricity taking it into the outer region of the solar system where its orbit could be modified by interactions with Uranus and Neptune. A close interaction between Triton and Pluto now gives the desired effect: Triton is captured in a retrograde orbit while Pluto is expelled into a heliocentric orbit with the orbital characteristics that we observe today. The feasibility of this hypothesis has been confirmed by computer modelling and the main conclusions of the model are given in Table 12.1.

Table 12.1 — Pre- and post-interaction orbital characteristics of Triton and Pluto with respect to the Sun and Neptune

	a(AU)	e
Before the interaction		
Triton–Sun	34.38	0.1633
Pluto–Neptune	0.0125	0.0
After the interaction		
Triton–Neptune (retrograde)	0.3212	0.9774
Pluto–Sun	39.52	0.2643

The initial orbit of Pluto would have been such that it repeatedly passed through the region of the interaction, through which Neptune's orbit also passed, so that there would be present the possibility of further interactions with Neptune. However, this condition would not long have prevailed. Because of Pluto's small mass and great distance from the sun, the resisting medium would not have been very effective in rounding off its orbit, but what *would* have been effective, particularly in view of the high orbital inclination, was the gravitational effect of the resisting medium causing precession. Over the course of time the differential rotation of the nodal lines would have separated the orbits of Neptune and Pluto so that, when the medium finally dispersed, the planets would have been completely isolated from each other, as indeed they are at present.

The computer calculations give a very eccentric orbit for the captured Triton, with eccentricity about 0.98. This would have been quite stable and subjected to a tidally mediated rounding mechanism which operates when a satellite is in orbit in the opposite sense to the spin of the parent planet (McCord, 1966). We also note that during the period of large eccentricity of Triton's orbit it would have been a considerable source of disturbance to any other satellite of Neptune, which could explain the large eccentricity of the orbit of Nereid.

A feature of our calculation which is worth noting is that it gives a nearest approach of Triton and Pluto of 3020 km, which is similar to the probable value of the sum of their radii. This suggests a very strong interaction between the bodies, perhaps even a collision which is not allowed for in our model but which could give a similar outcome. The presence of Charon can then be explained either in terms of tidal disruption of Pluto, an idea originally proposed by Harrington and Van Flandern (1979), or as due to break-up of Pluto by a direct collision.

A tidal origin of Charon was also suggested by Farinella *et al.* (1979) whose model has similar features to our own. They propose that Triton was captured from heliocentric orbit without the intervention of Pluto. Pluto was subsequently ejected from its orbit around Neptune by interaction with Triton after the orbit of Triton had sufficiently decayed by tidal friction. This again is quite feasible, although it would suggest a timescale so long that there would no longer be available a resisting medium to cause precession of Pluto's orbit and so isolate it from that of Neptune.

13

The origin of comets and meteorites

13.1 COMETS

Most comets have very elongated orbits, so before we discuss them in detail it will be useful to recall some results of solar-system dynamics. In Fig. 1.6 is shown a representation of an elliptic orbit with semi-major axis of length a, and eccentricity, e. The value of e for an elliptic orbit is constrained in the range zero to less than one, with zero corresponding to a circular orbit and one to a parabolic orbit. The aphelion distance, Q, is $a(1+e)$, and with very elongated orbits, for which $e \approx 1$, $Q \approx 2a$. The perihelion distance is $q=a(1-e)$.

A body in orbit around the sun possesses two components of energy: firstly kinetic energy due to the speed of its motion, which is a positive quantity, and, secondly, negative potential energy due to its distance from the sun. It is convenient to refer to the *intrinsic* energy of an orbiting body, E_i, by which is meant the energy per unit mass, and we may write

$$E_i = K_i + V_i \tag{13.1}$$

where K_i and V_i are the intrinsic kinetic and potential energies respectively. E_i is zero for a parabolic orbit, negative for an elliptic orbit and positive for a hyperbolic orbit, which is open and non-periodic (Fig. 6.1). For an elliptic orbit

$$E_i = -\frac{GM_\odot}{2a} \tag{13.2}$$

so that $1/a$ can be taken as a measure of intrinsic energy. In Fig. 13.1 there is reproduced a histogram by Marsden *et al.* (1978) giving the numbers of observed long-period comets (periods greater than 200 years) in intervals of $1/a$. The concentration around $1/a = 0$ had been noted by Oort (1948) who deduced that there existed a cloud of about 2×10^{11} comets at distances of tens of thousands of astronomical units. A comet can be observed only if it moves sufficiently close to the

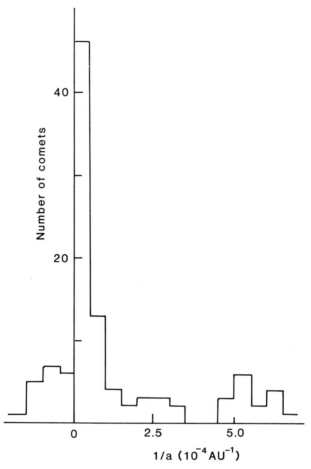

Fig. 13.1 — *Numbers of observed long-period comets in intervals of* 1/a. (After Marsden *et al.*, 1978).

sun — roughly within the orbit of Jupiter, although this varies with the physical characteristics of the comet itself. A comet with $1/a < 5 \times 10^{-5}$ AU will almost certainly be making its first trip into the inner solar system. The reason for saying this is that a comet entering the inner part of the solar system, where it is perturbed by the major planets, particularly Saturn and Jupiter, will acquire energy corresponding to a change of $1/a$ with average magnitude 8×10^{-4} AU^{-1}. If a change of such magnitude is added in a positive sense then the new energy will be positive and the comet will escape from the solar system; if the change is in a negative sense then this will produce a comet with a equal to a few thousand astronomical units, which is not the kind of orbit we are considering. We shall call these comets, which make a single excursion into the inner solar system with a some tens of thousands of astronomical units, *new* comets.

New comets are seen at the rate of about one per year. The inclinations of their orbits are random, and prograde and retrograde orbits are equally likely. The directions of the perihelion points have a bias towards the galactic plane (Tyror, 1957) and the average of the directions from which they come is about 23° from the apex of the sun's motion (Yabushita, 1979). Tyror (1957) also noted a tendency to have clusters of points (Fig. 13.2) for which, within each cluster, the comets tend to

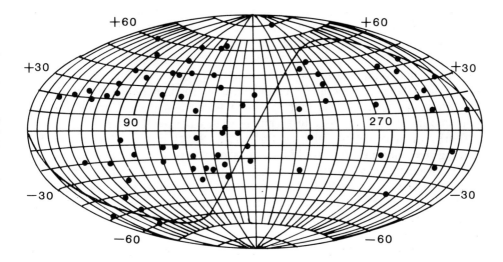

Fig. 13.2 — Directions of perihelion points of new comets. Some clustering of points may be seen.

have similar orbital characteristics.

The periods of new comets are several million years so that this is the minimum probable timescale for variations in their rate of arrival. Thus we cannot know whether the production of new comets is a steady-state phenomenon or whether we live in an unusual epoch with an abnormally small or large rate of arrival. With such constraints on our knowledge a large number of theories, based on different premises, have been advanced for the origin and evolution of comets.

The oldest of the ideas for perturbing the Oort cloud involves the influence of a star passing the solar system. The idea here is that the comet is on an elongated orbit with q greater than a distance at which major planet perturbation would be significant, but not too much greater. Perturbation by the star then changes the motion, mainly by subtracting angular momentum, so that when the comet moves in towards the sun its reduced perihelion distance renders it visible. For reasonable nearest approach distances of the star and relative sun–star velocities, the path of the star is approximately linear and the analysis of the change in q is straightforward. We have done such calculations, which agree well with those of Yabushita et al. (1982), and these are displayed in Table 13.1. It should be pointed out that the only plausible scenario for the production of a new comet is where the change of q goes from well outside the region of perturbation by major planets to some small value. Some thought will show that a model in which the comet moves closer to the sun with each successive orbit by small changes of q will not be tenable. The values in the table are for a perturbing star of mass equal to that of the sun moving past the solar system with a speed of 20 km s^{-1} acting on a comet with aphelion distance $40\,000$ AU. The changes of q would be amplified by having a more massive perturbing star, a smaller relative speed or a comet at a greater distance from the sun.

From knowledge of the density of stars in the galaxy and the distribution of the relative speeds of stars we have also calculated for each closest-approach distance, D_s, the mean time between such interactions (Table 13.1).

Given that the initial perihelion distance, q_0, should be greater than 20 AU, the table shows that stellar perturbation is an unlikely candidate for producing new comets unless there is one of the features given above for amplifying the perturbation — in which case one accepts that epochs in which new comets arrive are very spasmodic and the present time just happens to be a period of activity.

There is one more point which should be noted. If new comets were produced by a passing star, then the star should have passed by between one and three million years ago when the new comets were at or near aphelion. At a speed of 20 km s^{-1} the star would since have travelled a distance 40 ± 20 pc (1 parsec (pc) $\simeq 200\,000$ AU). Observations from the earth are capable of measuring both the distance and velocity (radial and transverse) of stars within 100 pc to sufficient accuracy to tell if a particular star *might* have passed close to the solar system within the last two million years or so. No such conclusion has ever been reported, although there are a few stars with small proper motions (corresponding to small velocities at right angles to the line of sight) which could be candidates.

More recently Clube and Napier (1984) have suggested that the Oort cloud could be very heavily affected by entities called giant molecular clouds (GMCs) which are detected in the galaxy by their CO emission. A GMC has a typical mass of $5 \times 10^5 M_\odot$ with radius 20 pc. It will have a hierarchical structure and contain smaller regions, which we shall call large molecular clouds (LMCs) of mass $2 \times 10^4 M_\odot$ and radius 2 pc which in their turn contain small molecular clouds (SMCs) of mass about $50_\odot$ and radius 0.08 pc. Clube and Napier have argued that the passage of the solar system through a GMC will sweep away the existing Oort cloud but, because the GMCs are rich sources of potential cometary material, the system will end up with a fresh supply of comets which it captures from the cloud.

Various calculations which have been done by ourselves and also by Bailey (1986) suggest that in the lifetime of the solar system the cometary family would have been considerably disturbed and partly lost by the action of GMCs, but that it is likely that most of the comets would be retained. In particular, virtually all the comets which originally had aphelia of $30\,000$ AU or less would be retained, although they would be greatly perturbed and, in particular, some would be transferred to orbits with aphelia of $50-80\,000$ AU in the main Oort cloud region. Bailey (1983) has suggested that there is an inner reservoir of comets much closer in than the standard Oort cloud and that the effect of GMCs is to remove some of the outer comets but also to

Table 13.1 — The maximum change of cometary perihelion due to a passing star of mass $M_\odot$ and various closest approach distances to the sun, D_m. The initial perihelion distance is q_0 and the final one, given in the body of the table is q_f. The sun–comet distance is 40 000 AU. The average time between interactions of the type specified is t_I.

	q_0(AU)				
D_s(pc)	20	15	10	8	t_I (years)
1.0	19.59	14.65	9.71	7.74	3.2×10^5
0.8	19.34	14.43	9.53	7.58	5.0×10^5
0.6	18.74	13.91	9.11	7.21	8.9×10^5
0.4	16.77	12.22	7.75	6.01	2.0×10^6
0.3	13.39	9.37	5.52	4.06	3.6×10^6

replenish the outer population by perturbation of the reservoir, as illustrated in Fig. 13.3 which comes from our own calculations.

Although GMCs are almost certainly significant from the standpoint of cometary evolution, they cannot be a steady-state influence on those bodies. Extreme assumptions show that the total number of fairly close interactions with GMCs in the lifetime of the solar system lies between one and ten, so that the intervals between such interactions must be at least 400 million years. This would lead to spectacular increases in new comet formation at such intervals but this would die out on the timescale of the cometary orbital periods.

Recently a much more promising mechanism has been proposed for giving a steady rate of arrival of new comets. This is based on the galactic tidal field and can be understood by reference to Fig. 13.4. Within the galactic disk the average star density falls off with increasing distance from the mean plane. For the point P there will be a net gravitational force towards the mean plane because there are more stars contributing to providing a field in that direction. The further away from the mean plane (within the disk), the higher is the net gravitational force, and this variation of force with distance constitutes the tidal field. Its dimensions are force per unit mass (acceleration) per unit distance and its estimated value in the vicinity of the solar system, allowing for the gravitational effect of 'dark matter' as equal to that of observable matter, is about 4×10^{-30} s^{-2}. This is only 40% of the tidal field due to a star of one solar mass at a distance of 1 pc, but the galactic field is more influential since it is always present whereas the tidal effects of a passing star may only be significant for 100 000 years or so.

We can do a fairly simple calculation to estimate the effect of the galactic field on a cometary orbit. In Fig. 13.5 there is shown a very elongated cometary orbit where the plane of the orbit is parallel to the direction of the galactic field gradient (giving maximum effect). The field at the sun is F_s and that at the comet is F_c so that the acceleration of the comet relative to the sun is

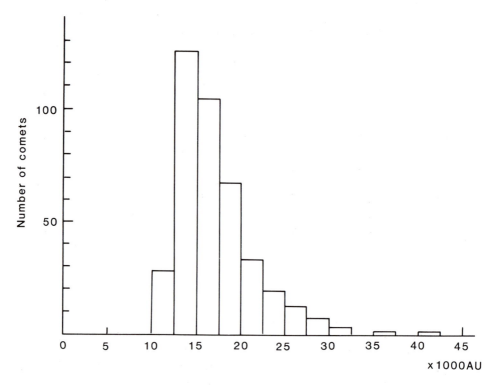

Fig. 13.3 — A histogram of values of a, originally all 15 000 AU, for 400 comets after a Monte-Carlo simulation of a solar-system journey through the galaxy. The duration of the journey was 4.5×10^9 years and only interactions with GMCs were considered. No comets were lost from the system.

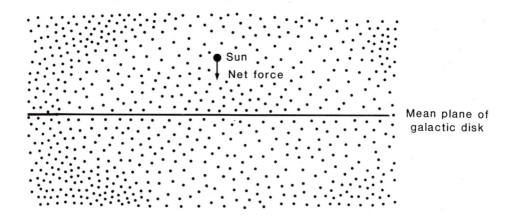

Fig. 13.4 — A representation of the distribution of stars in the galactic disk. The number density of stars falls off with distance from the mean plane.

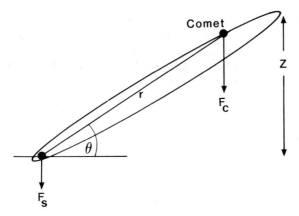

Fig. 13.5 — An elongated cometary orbit. F_s and F_c are the forces per unit mass (accelerations) due to the galactic tidal field at the sun and comet respectively.

$$F_c - F_s = Az = Ar\cos\theta, \tag{13.3}$$

where A is the galactic field gradient. This produces an intrinsic torque

$$T = Ar^2\cos\theta\sin\theta = \tfrac{1}{2}Ar^2\sin2\theta, \tag{13.4}$$

and if this was maintained for a time t then the intrinsic orbital angular momentum imparted to the comet would be

$$H = \tfrac{1}{2}Ar^2\sin2\theta t. \tag{13.5}$$

The intrinsic angular momentum of a body orbiting the sun is

$$H_0 = (GM_\odot p)^{1/2} \tag{13.6}$$

where p, the semi-latus rectum, can be replaced by $2q$ for an orbit where the eccentricity is close to unity.

We now assume that the comet spends about one-half of its orbital period close to aphelion and this contributes most of the change of angular momentum. This assumption gives a small but not very significant overestimate of the effect. The change in angular momentum as the perihelion changes from q_1 to q_2 is then given by

$$(2GM_\odot q_2)^{1/2} - (2GM_\odot q_1)^{1/2} = H . \tag{13.7}$$

Some results of calculation with this equation are given in Table 13.2 where H is taken as negative so that $q_2 < q_1$. They show that where the configuration of the orbit is favourable and with r greater than about 70 000 AU (i.e. $a > 35 000$ AU), the

Table 13.2 — The effect of the galactic tidal field on comets with various orbital characteristics. The initial perihelion distance is 15 AU. The final perihelion distance, given in the body of the table, assumes the most favourable inclination of the cometary orbit given the angle θ as defined in Fig. 13.5

	Semi-major axis (AU)			
$\theta(°)$	30 000	35 000	40 000	45 000
0 or 90	15.0	15.0	15.0	15.0
15 or 75	11.5	9.3	6.6	3.7
30 or 60	9.3	6.1	2.7	0.6
45	8.5	5.0	1.8	0.5

transformation from an orbit with perihelion outside the region of major planet perturbation to an orbit in which the comet will become visible is attainable. However, this does not convincingly explain all the observations of new comets. Their orbits include those which are not favourable for galactic tidal perturbation — for example with a less than about 20 000 AU and/or with orbits so positioned that galactic tidal effects would be small or even in the opposite direction to that required to produce a new comet. Again, the assumptions made about the amount of 'dark matter' in the galaxy have been challenged on the basis of new studies of galactic dynamics and the amount may be much less than is taken in the calculations above.

While the numerical results indicate that the galactic tidal effect is able to explain *some* new comets, it does not seem to be able to explain them all. In particular it gives no indication why there should be an observed clustering in the perihelion directions and why new comets coming from clusters tend to have similar orbital characteristics.

There is another scenario, based on the planetary-collision hypothesis, which could explain the characteristics of new comets. In the calculation, given in Chapter 10, which described the planetary collision, it was found that the more massive planet A was expelled from the solar system on a hyperbolic orbit of eccentricity just greater than unity. No consideration was given to the way in which planet A would have been affected structurally, although we have suggested that its outer volatile-rich layers in the region remote from the collison could have broken up and been separated from the main body of the planet by shock waves, thus forming a supply of potential cometary material. The orbits of such material would have had a range of variation from that of the planet; some would have had more energy and gone into even more hyperbolic orbits, but a proportion of it could have had less energy and gone into elliptical orbits of high eccentricity. The original orbits would have had perihelia in the terrestrial region of the solar system, but stellar and galactic field perturbations would, in time, have tended to increase the perihelia (Yasbushita *et al.*, 1982), so giving rise to a system like the Oort cloud. On this model, the existing comets would be a small fraction of the original volatile material in orbit, which would be consistent with there having been, say, between one and four earth masses of such material initially.

Now we consider another possibility: that the planet A was fragmented by the collision and that all or some of the fragments were retained by the solar system in very eccentric orbits. Precession and evolution of the orbits over the course of time would give a number of planetary-mass perturbers scattered throughout the comet cloud and we have calculated the effects that they would have on the comets themselves. In this calculation, the distribution of cometary orbits is somewhat different from that in the standard Oort model. This latter assumes a directionally isotropic distribution of velocities for comets at large distances from the sun, based on the randomizing effect of stellar and other perturbations. However, given original cometary orbits as indicated by the planetary-collision model, simulating stellar perturbations for the lifetime of the solar system shows a very non-isotropic distribution with a mean-square radial velocity up to six times that in a transverse direction. In Table 13.3 there is shown the expected number of new comets produced

Table 13.3 — The average rate of new comet formation per year for a planetary fragment of mass $100M_\oplus$ situated at different distances, D from the sun with 2×10^{11} comets in a non-isotropic distribution such that $\alpha = (v_{radial}^2/v_{transverse}^2)$. The fragment orbit has aphelion distance 23 000 AU and perihelion distance 100 AU

	D (AU)				
α	20 000	25 000	30 000	35 000	40 000
3	0.50	0.66	0.75	0.59	0.68
6	1.25	2.24	1.74	1.68	2.30

per year for a perturbing mass of $100M_\oplus$ for two different non-isotropic velocity distributions. The results were found by a Monte-Carlo process and hence fluctuate in a non-systematic manner. From our analysis it can be shown that the rate of new comet production depends on the square of planetary-fragment mass and hence, from the results in the table, it appears that four or five planetary fragments with total mass about $200M_\oplus$ could give the one new comet per year that is observed. While this mass is greater than that of planet A in Table 10.1, the increased mass can be accommodated without changing essential features of the planetary collision.

An additional feature of the planetary-fragment hypothesis is that the clumpiness found in the orbital directions can be explained. Very elongated orbits mean that for most of their paths the comets are moving in a direction making a small angle with the line of apses. The usual effect of a perturbation is to change slightly the direction of motion to produce an orbit with aphelion in the region of the perturber and direction of major axis very close to the original one. This means that new comets would tend to have their axis directions pointing towards the perturber, and that those influenced by a particular perturber would tend to have similar orbital characteristics. By the time that the comet became visible the orbit would have precessed due to the influence of galactic and stellar gravitational fields (about one radian in half an orbital period), and this would spread out the directions to give the kind of clustering observed by Tyror (1957) and shown in Fig. 13.2.

We conclude by summarizing the characteristics of the four mechanisms that have been suggested for perturbing the Oort cloud. GMCs are unable to account for steady-state new-comet formation because of the very long intervals between interactions with the solar system. On the other hand, when such interactions do occur they are likely to have a major effect, removing some of the more weakly bound comets but also replenishing the Oort population from an inner reservoir, if it exists.

Stellar perturbation is the least effective of the four processes, excepting for rare very-close encounters which may even match the action of a GMC. Like the GMCs they would give new comets spasmodically and could not give a steady-state formation rate. It should be said that it is possible, although unlikely, that we live in unusual times from the viewpoint of new comet formation. The problem then arises of why the most recent stellar perturber has not yet been identified.

Galactic tidal effects should be responsible for some, if not all, the new comets that are observed. We have seen that for a proposed perturbation mechanism to be effective it must be capable of moving the perihelion from 15 AU to less than 5 AU in a single orbit. If it is possible to identify all local (i.e. within the planetary region) perturbing influences with sufficient precision and also estimate the effects of stars for the last two million years or so, then it is possible, at least in principle, to integrate backwards the orbit of a new comet to see whether the galactic field would then transfer the perihelion to outside 15 AU. As far as is known, that exercise has not been attempted.

The effect of planetary perturbers within the Oort cloud is, by far, the most speculative of the four mechanisms; there is good evidence for the existence of GMCs, stars and a structured galactic disk. However, this idea has the advantage of explaining the pattern of observations rather better than do the other mechanisms and it derives from the idea of a planetary collision for which there is other circumstantial evidence.

13.2   METEORITES AND THEIR PROPERTIES

There is such a wealth of information available about meteorites that it is difficult to pick out those features which are of the greatest interest to the cosmogonist. Thus far we have only been concerned with the most obvious classification of meteorites in terms of their gross composition as irons, stony-irons and stones, and with the early thermal regimes they indicate. It is tempting to think of the different kinds of meteorites as components of a planet like the earth with the irons as representatives of metallic core material, stones as mantle or crust material and stony-irons as related to the interface between core and mantle. Here we shall consider in more detail than previously some of the physical characteristics of meteorites, but we shall also be describing other, more subtle, features before relating the properties of these bodies to the planetary-collision scenario.

As their name implies, the iron meteorites, which account for about six per cent of all falls, are mostly of that metal, but with a substantial admixture of nickel as well as lesser amounts of other elements — e.g. gallium, germanium and iridium. At high temperatures, above about 1200 K, a solid iron–nickel mixture of any composition will be in the form of taenite, an alloy of octahedral symmetry. However, on cooling, a nickel-poor Fe–Ni phase, kamacite, appears at the surface of the taenite crystals. The proportion of kamacite produced at any particular temperature depends on the proportion of nickel; with a low proportion of nickel a substantial conversion to kamacite occurs.

From a study of iron meteorites it is clear that they have been molten at some stage but then have cooled at a fairly slow rate, in the range 1–100 K per million years. The long timescale of cooling is necessary to explain the Widmanstätten figures (Plate 3). In the hot solid alloy there is a migration of nickel atoms to produce plates of kamacite between the taenite crystals; if the timescale for cooling is too short then nickel migration soon ceases and the pattern is not produced.

The stony meteorites are divided into chondrites and achondrites. The former type contain chondrules, small glassy silicate spherical inclusions, and these are by far the most common stones. The achondrites, which account for about eight per cent of falls and contain no chondrules, very often resemble the sort of igneously differentiated rocks found at the earth's surface.

On the basis of chemical composition, the chondrites are divided into five categories: E (enstatite), H (high iron), L (low iron), LL (low iron and low metal) and C (carbonaceous). The total amount of iron decreases as one goes from E to C and the degree of oxidation of iron is also graduated, so that enstatite chondrites contain mostly metallic iron whereas most of the iron in carbonaceous chondrites is in the oxidized state. There is also a trend that as the amount of oxidized iron increases so the iron : silicon ratio in the meteorite decreases.

Another type of classification depends on the distinctness of the chondrules in the meteorite. Types 2 and 3 show the chondrules sharply delineated against the background matrix (Plate 4) and the clarity with which the chondrules are seen decreases with increasing type number so that they are extremely indistinct in a type-6 chondrite. The type number 1 is used only for a variety of carbonaceous chondrite (C1) which actually contains very few chondrules at all.

An important property of the carbonaceous chondrites is that they contain an abundance of volatile elements — which are highly depleted in the other types of chondrite. There is considerable evidence of fractionation of elements in most chondrites. Elements may be classified into four classes — refractory (high boiling points), siderophile (metal loving), normal (moderate boiling points) and volatile. If a chondrite is found to be deficient in one element of a class, then all other elements of the same class will be depleted in unison. Again, for a given chemical group of chondrites the depletion of volatiles does seem to increase for increasing type number.

A great deal of work has been done in interpreting chondrite features in terms of condensation sequences from an initially hot but cooling solar nebula (section 3.5). The results of calculations have been encouraging, although the relationship to the temperature–pressure regimes which might actually occur in a solar nebula are not clearly established.

A final feature of meteorite composition, which has been attracting ever-increasing attention, is the existence of isotopic anomalies within them. As one example we consider oxygen with its three isotopes ^{16}O, ^{17}O and ^{18}O. Whatever the original relative abundances of isotopes for a particular element, there are various ways in which these abundances can change, e.g. by fractionation during a chemical reaction. However, in such a case the final composition will be related to the original one in a recognizable way since the degree of fractionation from one isotope to another is proportional to the difference of mass.

A device which is used to illustrate isotopic abundances in a case like oxygen is to plot isotope ratios. For oxygen the two used are:

$$\delta^{17} = \left\{ \frac{(^{17}O/^{16}O)_{sample}}{(^{17}O/^{16}O)_{SMOW}} - 1 \right\} \times 1000 \tag{13.8a}$$

and

$$\delta^{18} = \left\{ \frac{(^{18}O/^{16}O)_{sample}}{(^{18}O/^{16}O)_{SMOW}} - 1 \right\} \times 1000 \tag{13.8b}$$

where SMOW (standard mean ocean water) is an oxygen isotope standard.

If the sample is SMOW then clearly $\delta^{17} = \delta^{18} = 0$. However, if there is any chemical fractionation proportional to differences of mass number, then the change in δ values should give

$$\delta^{18} = 2\,\delta^{17} \tag{13.9}$$

It will be seen from Fig. 13.6 that most of the results from lunar, terrestrial and meteoritic material do lie either on the line given by equation (13.9) or on a line parallel to it. However, some material from C2 and C3 carbonaceous chondrites are found to give a line for which

$$\delta^{18} = \delta^{17} \tag{13.10}$$

and this cannot be explained in terms of chemical fractionation. It is, however, the line which would be obtained by adding pure ^{16}O in various amounts to some normal isotopic mixture and it has been suggested that pure ^{16}O was somehow injected into the solar system during, or shortly after, its formation.

Another isotope which has attracted a great deal of interest is ^{26}Mg (magnesium) which is the end-product of the radioactive isotope of aluminium, ^{26}Al, with a half-life of 7×10^5 years. The interest in this isotope is that if it was present in any quantity during the early stages of formation of the solar system then it would have been an important source of energy and even quite small bodies could have been melted. In some white, high-temperature inclusions in one carbonaceous chondrite, Allende, which contain a high ratio of aluminium to magnesium, it has been possible to measure excesses of ^{26}Mg with reasonable accuracy (Lee et al., 1976). It was found that the ^{26}Mg excess is proportional to the aluminium content and there is little doubt

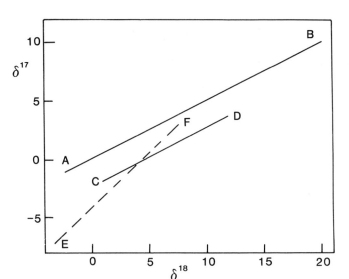

Fig. 13.6 — A δ^{17}–δ^{18} oxygen plot for: A–B terrestrial, lunar and differentiated meteorite samples; C–D matrix material from C2 carbonaceous meteorites; E–F anhydrous minerals in carbonaceous chondrites.

that in these samples the ratio $^{26}Al/^{27}Al$ was initially about 6×10^{-5}. However, the origin of the ^{26}Al and the extent of its presence in the early solar system is still an open question.

Finally we should look at neon, which has a common isotope ^{20}Ne and less common isotopes ^{21}Ne and ^{22}Ne. Normally the ratio $^{20}Ne/^{22}Ne$ is about 10, but there are some neon samples recovered from meteorites, so-called neon E, which are almost pure ^{22}Ne (Black and Pepin, 1969). The most likely interpretation of this is that the neon is the daughter product of radioactive ^{22}Na (sodium) which became trapped in cold meteorite rock when the sodium decayed. The problem here is that the half-life of ^{22}Na is about 2.6 years; if it is the source of ^{22}Ne then one needs a scenario where the radioactive sodium is produced in some nucleosynthetic reaction and becomes incorporated in rock which then becomes cool enough to trap the resultant neon, all on a timescale of a few years.

The account given here of meteorites and their properties has only scratched the surface of what has been discovered in the last two or three decades, but it is enough to show that no cosmogonist can ignore what the meteoriticist has to tell him. We have seen that the gross physical features of meteorites and their thermal history can readily be interpreted in terms of the outcome of a planetary collision. Now we shall see how well the observations of isotopic anomalies can be explained by the same event.

13.3   THE ORIGIN OF ISOTOPIC ANOMALIES IN METEORITES

Although there are many isotopic anomalies in meteorites, we shall confine our-selves to the three very significant ones: ^{16}O, ^{26}Mg and ^{22}Ne, the latter two being the daughter products of ^{26}Al and ^{22}Na respectively. The inferred presence of short-lived radioisotopes in the early solar system has led to the suggestion that there was a supernova event, or even a series of such events (Kirsten, 1978), preceding the formation of the solar system, or even initiating its formation in some way. In section 5.4 it was proposed that the injection of material by a supernova into the interstellar medium could lead to the formation of a dense interstellar cloud and hence trigger the formation of a stellar cluster. However, this would separate the supernova event from planetary formation by at least 10^7 years by which time many of the radioiso-topes would have disappeared. To counter this point, there are some who argue that the newly formed radioisotopes were trapped in small solid grains, which were subsequently incorporated into solar system material in a cold form that preserved the record of the original composition. Such an argument is tenable except, perhaps, for ^{22}Na; it is not at all obvious that the time to go from formation within the supernova to being incorporated in a cold solid rock, which will retain released ^{22}Ne, could be as small as a few times the half-life of ^{22}Na (2.6 years). Nevertheless, what seems to be beyond dispute is that there was some radiosynthetic event associated with the origin of the solar system and we should examine the reactions which could give what is observed.

A reaction which can produce ^{16}O is

$$^{12}C + {}^4He \rightarrow {}^{16}O \tag{13.11}$$

requiring the presence of carbon and helium in some abundance at a temperature somewhat above 10^8 K. At similar temperatures, other reactions can give ^{22}Na and ^{26}Al; these are

$$^{22}Ne(p,n)^{22}Na \tag{13.12}$$

and

$$^{26}Mg(p,n)^{26}Al \tag{13.13}$$

The interpretation of these reaction formulae is that, for example, a proton reacts with ^{22}Ne to give a neutron and ^{22}Na.

In this scenario we require a typical cosmic mix of materials at high temperature, and preferably high density, in which the above reactions can take place. When the material cools to the temperature at which an aluminium-containing mineral can form, then some small part of the aluminium will be ^{26}Al. If the mineral also contains little magnesium (which may or may not have a reduction of ^{26}Mg depending on its source) then an excess of ^{26}Mg will be recorded.

As we have seen, it is ^{22}Na which is the major constraint on theories because of the short timescale it imposes. For that reason we are going to examine an alternative scenario which takes place within the newly formed solar system itself, where cooling timescales are much shorter and densities much higher than in the vicinity of a supernova.

The scenario we have in mind involves the postulated planetary collision, which has already provided explanations for many other features of the solar system. A planet in the asteroid-belt region, moving on an orbit of fairly high eccentricity, would have an orbital speed of about 25 km s^{-1}, and two such planets, with suitable orbits, could have an approach speed of 40 km s^{-1} or so. The actual speed of the collision would be greater than this since they attract each other. Two bodies of masses M_1 and M_2 with radii r_1 and r_2 coming together from zero relative speed at large distance will, when they just touch, have a speed V_ϕ given by

$$V_\phi^2 = 2G(M_1 + M_2)/(r_1 + r_2) , \qquad (13.14)$$

and if the speed at large distance apart is V_∞ then the speed of collision, V_c is given by

$$V_c^2 = V_\phi^2 + V_\infty^2 . \qquad (13.15)$$

With $V_\infty = 40$ km s^{-1}, $M_1 = 20M_\phi$ $M_2 = 100M_\phi$, $r_1 = 20\,000$ km and $r_2 = 30\,000$ km, we find $V_c = 58$ km s^{-1}.

The amount of energy available for conversion into thermal energy is somewhat higher than this because the material is still accelerating inwards after the planets touch. Some of the energy will be retained as kinetic energy in ejecta material, some, in fact a considerable fraction, will be absorbed in ionizing the material in the region of impact — that is to say stripping electrons off atoms — while the rest will contribute to raising the temperature of the material. Ionization acts in two ways to reduce the temperature; the first is that energy is required to pull the electrons away from the atoms, and the other way is that it produces more particles to share the thermal energy.

The main area of heating is in the collision region. A mathematical model of the collision has been investigated in which the impact region has been taken to consist of a mixture of hydrogen, ices and silicates, as might be expected in the atmospheres and on the surface of major planets. Depending on the assumptions put into the modelling, the temperatures reached are usually in the range 2–4 × 10^6 K — high indeed, but nowhere near high enough to give nuclear reactions which could produce ^{16}O, $^{26'}$Al and ^{22}Na.

To pursue this matter further we need to consider the way in which planets retain atmospheres and the mechanisms for the evolution of atmospheres on newly formed planets. Atoms or molecules of mass μ at a temperature θ have a distribution of speeds (the Maxwell distribution) with a root-mean-square speed given by

$$v_\theta^2 = 3k\theta/\mu . \qquad (13.16)$$

If a molecule in the upper reaches of a planetary atmosphere, where collisions are rare, attains a speed greater than the excape speed from the planet then it will be lost. In Fig. 13.7 there is shown a Maxwellian distribution of speeds for a gas. If the escape speed from the planet is v_{e1} then the atmosphere will be lost on a very short timescale; if it is v_{e2} the atmosphere may be retained for a long time, say 10^6 years;

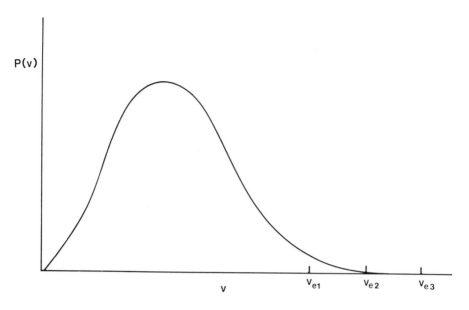

Fig. 13.7 — The Maxwell distribution of speeds for gas molecules of a particular kind in a planetary atmosphere. Comparatively small differences in escape speed from the planet can give very large differences in the probability of escape of molecules.

while if the escape speed is v_{e3} then the atmosphere will be retained indefinitely. When the theory of atmospheric retention or loss is done in detail, it turns out that the factor which determines what happens is the ratio ψ, given in equation (12.1). In

Table 13.4 — The ratio $\psi = v_e^2/v_\theta^2$ for a number of solar system bodies and various actual or possible atmospheric gases

Planet or satellite	Gas	Radius (km)	θ (K)	Mass ($M_\oplus$)	ψ	Type of atmosphere
Jupiter	H_2	71 000	120	318	594	Thick
Earth	N_2	6400	295	1	118	Moderate
Mars	CO_2	3400	250	0.108	44.5	Thin
Titan	N_2	2575	90	0.024	22.8	Moderate
Mercury	CO_2	2400	620	0.055	12.9	None
Moon	O_2	1740	290	0.012	6.2	None

Table 13.4 there are shown the values of ψ for a number of solar-system bodies, some of which have atmospheres and some of which do not. It is clear that there is a critical value of ψ, about 20, below which no atmosphere can exist on a long timescale and above which atmospheres are actually observed.

There are processes by which atmospheric constitutents may be lost even though ψ may be above the critical value. For example, molecules may be dissociated by the action of ultraviolet light in the upper reaches of the atmosphere so that, for example, H_2O may be dissociated into OH and H and the hydrogen may be lost. Such a mechanism is thought to be responsible for the loss of most of the water on Venus. Again, early planets would certainly have been much warmer than just their position in the solar system would have indicated since they would have contained considerable thermal energy due to collapse to planetary density. For example, the heat due to the accretion of a body the mass and size of the earth would have taken about one million years to radiate away if its surface was at a constant temperature of 1000 K.

We now consider an early protoplanet, such as the less massive of the ones that collided, in a condition where hydrogen atoms, produced by dissociation of molecules in the upper atmosphere, have $\psi = 10$, say. All hydrogen contains a small proportion of the stable isotope, deuterium (D), which in Jupiter, interstellar space and the cosmos at large gives a D/H ratio of 2×10^{-5}. However, large variations in this ratio occur. For the earth it is 1.6×10^{-4}, for some meteorites more than six times the terrestrial value, while for Venus it is 1.6×10^{-2}, 100 times the value for the earth. If the atmosphere of the early protoplanet contained H_2, H_2O and NH_3, then it will also have contained HD, HDO and NH_2D. Dissociation which yielded hydrogen would have led to its loss, while any deuterium produced would have been retained sufficiently long for it to be reincorporated into the atmospheric components. Given that the lifetime of the molecular hydrogen component of the atmosphere was long, but not infinite, a great deal of its deuterium content would have been transferred into the other molecular species present. When the protoplanet eventually cooled, so the residual atmospheric gases would have condensed to form a layer of ices of water, carbon dioxide, ammonia and perhaps methane on the planetary surface, and this layer would have been much enriched in deuterium.

What we have now established is that a planetary collision could create a temperature of $2–4 \times 10^6$ K in planetary surface material highly enriched in deuterium, and this is just the condition to set off a thermonuclear reaction. At such temperatures deuterium undergoes the reactions

$$D(D,n)^3He \quad \text{energy per reaction} \quad 3.27 \text{ MeV} \tag{13.17}$$

$$D(D,p)^3T \quad \text{energy per reaction} \quad 4.03 \text{ MeV} \tag{13.18}$$

where the two branches, of approximately equal probability, are referred to as the neutron and proton branches respectively. In the second reaction T is tritium, a radioactive isotope of hydrogen. The tritium itself reacts with deuterium very readily, thus

$$D(T,n)^4He \quad \text{energy per reaction} \quad 17.6 \, MeV \tag{13.19}$$

while a reaction involving ^3He that is slower and less efficient at the lower temperatures of interest but contributes considerable energy when the temperature rises, is

$$D(^3He,p)^4He \quad \text{energy per reaction} \quad 18.3 \, MeV. \tag{13.20}$$

Calculations have been carried out in which surface planetary material consisting of a mixture of silicates and ices, with a D/H ratio of 1.4×10^{-2}, less than the Venus value, is subjected to triggering temperatures of $2 \times 10^6 \, K$ and $3 \times 10^6 \, K$. In both cases thermonuclear reactions give final temperatures of order $2 \times 10^8 \, K$, the timescale being 1700 s for the lower temperature and approximately 1 s for the higher one. The actual time for which the colliding planets will continue to move together after the initial contact, and so maintain the conditions for the reaction, is of the order of the time required to traverse a planet at the speed of the collision, which is a few hundred seconds.

With helium and some hydrogen present as part of the residual atmosphere, which would be expected for a fairly massive planet, all the ingredients and the conditions would be available to give the reactions described in equations (13.11)–(13.13). The vaporized material would expand and cool to temperatures at which solid grains could form, with a density high enough to ensure that the grains were produced quickly and then be rapidly incorporated into chondritic meteorites — as is required to explain the Neon E observations.

It can be seen that a natural sequence of events, related to the planetary-collision hypothesis, accounts for all the important features of meteorites, even the ones which impose the tightest constraints on theory.

14

What the capture theory explains

14.1 PATTERN AND ORDER

In Chapter 4 there was a given list of features of the solar system that it would be either necessary or desirable for a theory to explain. However, it would not be enough just to be able to give an explanation for every feature. To be truly plausible a theory which explains various aspects of the origin, evolution and present structure of the solar system should itself have a desirable structure. If large numbers of solar system features are given explanations which relate to no other feature, then the theory would be disjointed and unsatisfactory; while most investigators would accept that the solar system has experienced many irreversible events, that is not to say that these events are all random and disconnected. We would assert that a plausible theory is one in which strong connections do exist, linking different features of the solar system, and where there is a causal pattern linking each event to some previous happening. Our claim is that origin via the Capture Theory, the planetary-collision hypothesis and the consequences of the collision do provide the connected causality which is desirable, and we now summarize what has been described in previous chapters to bring out this pattern.

14.2 THE INTERSTELLAR MEDIUM AND STAR FORMATION

Our model begins with interstellar space, sparsely populated with hydrogen plus a small 1–2% component of heavier atoms and with a density around 10^{-21} kg m^{-3} and a temperature of order 10 000 K. The universe has existed already for some 10^{10} years and stars are present in all the stages of their evolution, from birth to death, that are observed today. The trigger for all that follows is a supernova event which injected heavy elements, much of it in the form of fine dust and grains, into a surrounding medium. The modified interstellar medium was then no longer in thermal equilibrium with its surroundings since it had a higher density of coolants, so it fell in temperature, thus reducing the pressure and so sucking in surrounding material to increase the pressure to the surrounding ambient level. Eventually a state was reached where there was a dense, cool cloud in both thermal and pressure equilibrium with the surrounding interstellar medium.

While thermal and pressure equilibrium has been attained there was another factor at play and that was the mass of the interstellar cloud. This was greater than the Jeans' critical mass and so the cloud began to collapse, slowly, almost imperceptibly, at first but gathering momentum and generating violently turbulent internal motions as the collapse progressed. Eventually the stage was reached where colliding turbulent elements formed hot, high-density regions within the cloud. The rate at which these cooled was much faster than the rate at which they expanded, so that they became dense cool regions able to collapse individually. When one of these regions collapsed to high density before the overall turbulence of the cloud stirred its material back into the background material, a star formed. In this way the collapsing cloud produced a galactic cluster of between 100 and 1000 stars. The way of forming the stars, by collision of turbulent streams, ensured that they spun slowly, although stars gaining matter by accretion did pick up the angular momentum associated with the more massive so-called early-type stars. The angular momentum of the original region of interstellar space was conserved, but existed in the motion of the stars within the cluster rather than in the spins of individual stars.

14.3 INTERACTIONS BETWEEN STARS

In the newly forming galactic cluster the stars would have been relatively close together and would have been at different stages of development. Stars of solar mass would have existed in the company of other less-massive stars that were still fairly diffuse objects collapsing towards the main sequence. Interactions between stars would have been happening at this stage. In some cases, pairs of stars coming close together, with other bodies nearby to take up energy, would form binary systems, and existing binary systems could even be broken up by perturbation effects in the many-body environment.

Other kinds of interaction, involving tidal effects on diffuse bodies, also occurred, the outcome depending on the parameters of the interaction. For example, some would have given partial disruption of the diffuse star which left some of its substance forming an envelope around the residual central body. When the central body collapsed there would exist a forming star surrounded by a nebula-like cloud of gas and dust, as has been recorded by infra-red detection in recent years. However, as theory shows, while such a diffuse cloud could give rise to small condensed bodies going around the central star, large bodies, resembling major planets, would not form.

The interaction which gave rise to the solar system was a close passage involving the embryonic sun, a condensed body but not necessarily on the main sequence, with a collapsing diffuse protostar, with perhaps a quarter of a solar mass. As the protostar passed close to the sun, so a tidal filament was drawn from it, and within this filament, by a mechanism described by Jeans, six individual condensations formed. Because of the nature of the interaction, these condensations were captured into solar orbits with initial motions taking them to large distances, many tens of astronomical units, from the sun. For this reason, before the planets returned to their perihelion regions, they had time to condense sufficiently to resist solar-tidal disruption. However, they were subjected to strong tidal effects that imparted considerable amounts of angular momentum into tidal bulges. In the final stages of planetary collapse, the exaggerated bulges, in the form of filaments, broke up to give families of natural satellites. Much of the tidal bulge material was eventually subsumed into the main body of the planet, so giving rise to the planetary spins.

14.4   EVOLUTION OF ORBITS

The original orbits of the planets were elongated with inclinations up to $10°$ from the mean plane of the system. Around the sun was a resisting medium consisting of gas and dust pulled out of the protostar, some of which had never been part of the tidal filament but other parts of which was filament material which had not gone into planetary condensations. With a total of about five Jupiter masses, this resisting medium would have had a lifetime of some 10^8 years, the gaseous component being largely expelled from the solar system by radiation pressure while the larger solid grains were sucked into the sun by the Poynting–Robertson effect. The angular momentum of this latter material, corresponding in direction to rotation in the mean plane of the system, combined with the original angular momentum of the sun gave the tilt of the spin axis of the sun at $7°$ to the normal to the mean plane.

The planets underwent a process of rounding off with timescales heavily dependent on their masses. Thus Jupiter took only 10^5 years to round off, while Uranus took about 2×10^6 years. The two innermost planets, with masses now estimated to be about $200 M_\oplus$ and $25 M_\oplus$, were in orbits which would have rounded off to radii of 2.8 and 1.6 AU on timescales in the range 1–5×10^6 years. The revised estimates of planetary masses, differing from those given in Table 10.1, are based on the considerations mentioned in section 13.1.

There were characteristics other than just round-off in the evolution of the orbits. The resisting medium had a form in which it was flattened along the axis of rotation and this gave a gradual, but not montonic, reduction of the original inclination of the orbit. In addition, the resisting medium exerted a small gravitational force on the planets and, because of its flattened form, this force was non-central, i.e. not pointing towards the sun and with a component out of the plane of the orbits. The effect of this was to cause the orbits to precess; that is to say that there was a rotation of the line of apses joining perihelion to aphelion while simultaneously and non-synchronously there was a rotation of the line of nodes, the line in which the planetary orbit crossed the mean plane of the system.

The scenario now created is of a system of planets, with attendant regular satellites, in orbits with inclinations not more than $10°$ from the mean plane but gradually decreasing and with differential precessions which caused pairs of orbits to cut through each other from time to time. The conditions for interactions between planets now existed.

14.5   INTERACTIONS BETWEEN PLANETS

The early protoplanets would have collapsed to density high enough to resist disruption by tidal effects at perihelion passage on a timescale of a few tens of years. However, during their first one or two orbits they could be substantially influenced by close passage to other planets. For example, the passage of Jupiter at a distance of 0.03 AU (2.4 times the orbital radius of Callisto) would have been equivalent to a solar passage at distance 0.3 AU, which would give considerable tidal effects. This is the mechanism by which the spin axes of the planets were given their present directions. That of Uranus was caused by a very close passage such that the vector of nearest approach was almost perpendicular to its orbit giving a spin slightly retrograde with an axis close to the orbital plane.

For pre-existing satellites these close planetary interactions would have been a source of considerable disturbance, certainly destroying the pattern of circular orbits in the equatorial plane. The process of planetary collapse in the presence of tidal influences gave an accompanying halo of material, denser than the prevailing resisting medium in the solar system as a whole, which acted as a local resisting medium accompanying each planet. The combined presence of this, and the gravitational forces due to the oblate-spheroidal shape of the rapidly spinning major protoplanets, quickly restored the pattern of regular satellites.

After a million years, more or less, there were six planets of which only Jupiter and Saturn had attained fairly circular orbits. The outermost planet, Neptune, had only two small satellites since it had been subjected to the weakest solar tidal effects, although a subsequent planetary interaction had tilted its spin axis by some 29°. The two innermost planets, A and B, on the other hand had been subjected to very large solar tidal effects and had some very large satellites, including one four times as massive as Ganymede.

The planetary orbits evolved and precessed but, as mathematical analysis shows, as time passed so the cumulative probability of some major event taking place increased to the point where some such event was more likely than not. Eventually the planets A and B collided.

14.6   THE EFFECTS OF A PLANETARY COLLISION

It is conjectured that one of the planets which collided had a mass between that of Saturn and Jupiter while the other, the innermost one, had a mass nearly twice that of Neptune. The larger one consisted of a solid iron–silicate core covered by ice-impregnated silicates and ices, together with a dense atmosphere of hydrogen and hydrogen-containing compounds. Indeed, conditions may have been such as to produce regions of metallic hydrogen, such as have been postulated for Jupiter. The inner planet lacked the massive content of hydrogen, although it might have had a fairly dense methane plus hydrogen atmosphere, most other potential atmospheric components having condensed to form an icy mantle when the planet had sufficiently cooled. The D/H ratio in the hydrogen content of the outer regions was similar to that of Venus (around 1.6×10^{-2}) since hydrogen atoms formed by dissociation of molecules in the upper atmosphere had been lost while deuterium was retained.

The primary effect of the collision was to break up the smaller planet into two major and one minor component which lost energy and fell into the inner part of the solar system. In this environment the volatiles were not retained and, after rounding-off of orbits (the resisting medium was still around and was reinforced somewhat by the collision debris), these bodies formed the earth, Venus and Mercury. The more massive planet gained energy from the collision and fragments of it were sent into orbits of large eccentricity and semi-major axes of length many thousands of astronomical units. These fragments were accompanied by a myriad of small ice-silicate fragments, of kilometre dimensions.

The satellites of the colliding planets survived the collision in a variety of ways. Some accompanied the fragments into large scale orbits but others were retained within the confines of the observable solar system. One very large satellite, the present planet Mars, went into an independent heliocentric orbit which rounded off close to, but slightly within, the orbit which would have been taken up by the inner planet. Another was captured by the earth-forming fragment of the inner planet and exists today as the moon. Both Mars and the moon were left with hemispherically asymmetric surface topography due to their former proximity to the planetary collision.

Another large satellite, identified as Triton, originally in a heliocentre orbit taking it out just beyond the orbit of Neptune, eventually passed so close to Neptune that it had a close interaction with its larger satellite, Pluto. The effect of this was to retain Triton as a satellite of Neptune in a retrograde orbit and to eject Pluto into a heliocentric orbit which has the characteristics we see today. The separation of the two orbits came about by the precession of Pluto's orbit due to the effect of the residual resisting medium. The close passage of Triton and Pluto caused tidal disruption of the latter body and hence its satellite companion, Charon.

In the period following the planetary collision, the debris would have repeatedly moved into the solar system causing the collision damage on various bodies that is still visible today. Collisions within the gravitational influence of major planets would have given, for example, the two families of irregular satellites of Jupiter. However, the effect of perturbations by passing stars, the galactic tide, the planetary fragments and by occasional interactions with a giant molecular cloud (GMC) would eventually have moved the perihelia of much of the debris beyond the orbits of the known planets. Of the several earth masses of original debris, some one-tenth to one-hundredth of an earth mass remains to provide the comets of which we see evidence today. The inner reservoir, with aphelia of order 10^4 AU, when perturbed by a GMC, provides reinforcements for the Oort cloud, the original members of which are somewhat ravaged by the same GMC. The new comets seen today are produced by both galactic tidal effects and also occasional perturbation with planetary fragments, which gives the observed clustering of perihelion directions and similarities of orbital characteristics.

The observed characteristics of meteorites can also be explained in terms of a planetary collision. The major division of meteorites into irons and stones came from density separation of material in hot massive planets. In the various temperature regimes which existed immediately following the planetary collision, there occurred chondrule formation, both by the splashing of molten silicates and by droplets condensing from vaporized material. Depending on their mode of formation and their rate of cooling when associated with matrix material, the chondrules would be more or less distinct in the final cold meteorite, so giving rise to different classes of chondrites. Thus a body, or part of a body, cooling slowly would produce a chondrite with a higher type number (i.e. with less distinct chondrules) while at the same time it would lose more of its volatile inventory. This is the association noted in section 13.2. Silicate material which escaped major heating and survives in its original form without admixture with collision products forms the rarer, metal-poor achondrites.

The planetary collision directly produced temperatures in the collision region which were capable of igniting deuterium-based reactions. This produced local temperatures of order 10^8 K and hence nuclear reactions explaining observations of extinct ^{26}Al and ^{22}Na and present excesses of ^{16}O.

14.7 DOES THE CAPTURE THEORY EXPLAIN EVERYTHING?

In Chapter 4 there were outlined the desirable features of what should be explained by a plausible theory of the origin and evolution of the solar system. The Capture Theory seems to deal satisfactorily with all of these, although it may be argued that point (4), dealing with the predicted frequency of planetary systems, is not convincingly shown. Nor indeed has the process of formation of natural satellites been well modelled in detail, although suggestive calculations have been done to support the proposed mechanism.

Despite these minor shortcomings, which are of omission rather than contradiction, the Capture Theory does provide a model in terms of which various features fall into place as part of the evolving pattern of the system as a whole. The only other theory for which this might be true is McCrea's floccule theory which would be a strong one if it could overcome the objections raised in Chapter 3.

However, the most important fact to realize in work of this kind is that *we can never know*. Any theory, no matter how convincing it may seem at any particular time, may be negated by new evidence with which it is totally incapable of being reconciled. Any theory worth taking seriously must be vulnerable. The Capture Theory is certainly vulnerable and, for the time being at least, is worth taking seriously.

References

Abt, H. A. (1977) *Scient. Amer.* **236** (4) 96–104.

Alfvén, H. (1978) *The origin of the solar system.* (ed. S. F. Dermott), pp. 41–48. Wiley: Chichester.

Alfvén H. and Arrhenius, G. (1975) *Structure and evolutionary history of the solar system.* D. Reidel: Dordrecht.

Allen, C. C. (1979) *Icarus* **39** 111–123.

Allison, D. J. (1986) *The early evolution of a protoplanet.* Ph.D. Thesis, CNAA.

Anders, E. (1971) *Physical studies of minor planets.* (ed. T. Gehrels), pp. 429–446. NASA SP-267, Washington, D.C.

Anders, E. and Owen, T. (1977) *Science* **198** 453–465.

Arvidson, R. E., Goettel, K. A. and Hohenberg, C. M. (1980) *Rev. Geophys. Space Science* **18** 565–603.

Aust, C. and Woolfson, M. M. (1971) *Mon. Not. R. Astr. Soc.* **153** 21P–25P.

Aust, C. and Woolfson, M. M. (1973) *Mon. Not. R. Astr. Soc.* **161** 7–13.

Bailey, M. E. (1983) *Mon. Not. R. Astr. Soc.* **104** 603–633.

Bailey, M. E. (1986) *Mon. Not. R. Astr. Soc.* **218** 1–30.

Benz, W., Slattery, W. L., and Cameron, A. G. W. (1986) *Icarus* **66**, 515–535.

Benz, W., Slattery, W. L., and Cameron, A. G. W. (1987) *Icarus* **71** 30–45.

Berlage, H. P. (1930) *Proc. K. Ned. Akad. Wet.* **33** 614–618.

Black, D. C. and Pepin, R. O. (1969) *Earth Planet. Sci. Lett.* **6** 395–406.

Bondi, H. (1952) *Mon. Not. R. Astr. Soc.* **112** 195–204.

Bondi, H. and Hoyle, F. (1944) *Mon. Not. R. Astr. Soc.* **104** 273–282.

Buffon. G. L. L. (1745) De la formation des planètes. In *Oeuvres complets* (1852), Vol. 1, pp. 120–140. Brussels: Adolf Deros.

Cameron, A. G. W. (1978) *The origin of the solar system.* (ed. S. F. Dermott), pp. 49–74. Wiley: Chichester.

Chamberlin, T. C. (1901) *Astrophys. J.* **14** 17–40.

Clube, S. V. M. and Napier, W. M. (1984) *Mon. Not. R. Astr. Soc.* **208** 575–588.

Coates, I. E. (1980) *On the origin of planets.* Ph.D. Thesis, CNAA.

Connell, A. J. and Woolfson, M. M. (1983) *Mon. Not. R. Astr. Soc.* **204** 1221–1230.

Disney, M. J., McNally, D., and Wright, A. E. (1969) *Mon. Not. R. Astr. Soc.* **146** 123–160.

Dormand, J. R. and Woolfson, M. M. (1971) *Mon. Not. R. Astr. Soc.* **151** 307–331.

Dormand, J. R. and Woolfson, M. M. (1974) *Proc. R. Soc. Lond.* **A340** 307–331.

Dormand, J. R. and Woolfson, M. M. (1977) *Mon. Not. R. Astr. Soc.* **180** 243–279.

Dormand, J. R. and Woolfson, M. M. (1980) *Mon. Not. R. Astr. Soc.* **193** 171–174.

Dormand, J. R. and Woolfson, M. M. (1988) *The physics of the planets.* (ed. S. K. Runcorn), pp. 371–383. Wiley: Chichester.

Eddington, A. S. (1926) *The internal constitution of the stars.* Cambridge University Press.

Farinella, P., Milani, A., Nobili, A. M., and Valsacchi, G. B. (1979) *Moon Planets* **20** 415–421.

Freeman, J. W. (1978) *The origin of the solar system.* (ed. S. F. Dermott), pp. 635–640. Wiley: Chichester.

Gault, D. E. and Heitowit, E. and D. (1963) *Proc. Sixth Hypervelocity Impact Symp.,* Cleveland, Ohio, April 30–May 2.

Gingold, R. A. and Monaghan, J. J. (1977) *Mon. Not. R. Astr. Soc.* **181** 375–389.

Goldreich, P. and Ward, W. R. (1973) *Astrophys. J.* **183** 1051–1061.

Guest, J., Butterworth, P., Murray, J., and O'Donnell, W. (1979) *Planetary geology.* David & Charles: London.

Harrington, R. S. and Van Flandern, T. C. (1979) *Icarus* **39** 131–136.

Hayashi, C. (1961) *Publs. Astr. Soc. Japan* **13** 450–452.

Hayashi, C. (1966) *A. Rev. Astron. & Astrophys.* **4** 171–192.

Herbig, G. H. (1978) *The origin of the solar system* (ed. S. F. Dermott), pp. 219–236. Wiley: Chichester.

Hoyle, F. (1960) *Q. J. R. Astr. Soc.* **1** 28–55.

Hutchison, R., Alexander, C. M. O., and Barber, D. J. (1988) *Phil. Trans. R. Soc. Lond.* **A325** 445–458.

Iben, I. and Talbot, R. J. (1966) *Astrophys. J.* **144** 968–977.

Jeans, J. H. (1917a) *Mon. Not. R. Astr. Soc.* **77** 186–199.

Jeans, J. H. (1917b) *Mem. R. Astr. Soc.* **62** 1–48.

Jeans, J. H. (1919) *Problems of cosmogony and stellar dynamics.* Cambridge University Press.

Jeffreys, H. (1918) *Mon. Not. R. Astr. Soc.* **78** 424–442.

Jeffreys, H. (1929) *Mon. Not. R. Astr. Soc.* **89** 636–641.

Jeffreys, H. (1952) *Proc. R. Soc. Lond.* **A214** 281–291.

Kirsten, T. (1978) *The origin of the solar system.* (ed. S. F. Dermott), pp. 267–346. Wiley: Chichester.

Lamy, P. L. and Burns, J. A. (1972) *Amer. J. Phys.,* **40** 441–445.

Laplace, P. S. de (1796) *Exposition du système du monde.* Paris: Imprimerie Cercle-Social.

Lee, T., Papanastassiou, D. A., and Wasserburg, G. J. (1976) *Geophys. Res. Lett.* **3** 109–112.

Lynden-Bell, D. and Pringle, J. E. (1964) *Mon. Not. R. Astr. Soc.* **168** 603–637.

Lyttleton, R. A. (1960) *Mon. Not. R. Astr. Soc.* **121** 551–569.

Lyttleton, R. A. (1961) *Mon. Not. R. Astr. Soc.* **122** 399–407.

Lyttleton, R. A. (1972) *Mon. Not. R. Astr. Soc.* **158** 243–279.

Marsden, B. G., Sekanina, Z. and Everhart, E. (1978) *Astr. J.* **83** 64–71.

McAdoo, D. and Burns, J. A. (1975) *Earth Planet. Sci. Lett.* **25** 347–354.

McCord, T. B. (1966) *Astr. J.* **71** 585–590.

McCrea, W. H. (1960) *Proc. R. Soc. Lond.* **A250** 245–265.

McCrea, W. H. (1978) *The origin of the solar system* (ed. S. F. Dermott), pp. 75–110. Wiley: Chichester.

McElroy, M. B., Ten Ying Kong, and Yuk Ling Yung (1977) *J. Geophys. Rev.* **82** 4379–4388.

Moulton, F. R. (1905) *Astrophys. J.* **22** 165–181.

Murray, B. C. and Mallin, M. C. (1973) *Science* **179** 997–1000.

Napier, W. McD. and Dodd, R. J. (1973) *Nature* **242** 250–251.

Oort, J. H. (1948) *Bull. Astr. Insts. Neth.* **11** 91–110.

Owen, T. and Bieman, K. (1976) *Science* **193** 801–803.

Pollack, J. B., Kasting, J. F., Richardson, S. M., and Poliakoff, K. (1987) *Icarus* **71** 203–224.

Roche, E. (1854) *Mem. Acad. Montpellier* **2** 399–439.

Runcorn, S. K. (1975) *Phys. Earth and Planet. Int.* **10** 327–335.

Runcorn, S. K. (1980) *Proc. Lunar Planet. Sci. 11th Conf.* p. 1867.

Russell, H. N. (1935) *The solar system and its origin.* MacMillan: New York.

Safronov, V. S. (1972) *Evolution of the protoplanetary cloud and formation of the earth and planets.* Israel Program for Scientific Translations: Jerusalem.

Schmidt, O. Y. (1944) *Dokl. Akad. Nauk USSR,* **45** No. 6.

Schmidt, O. Y. (1959) *A theory of the origin of the earth; four lectures.* Lawrence & Wishart: London.

Schofield, N. and Woolfson, M. M. (1982a) *Mon. Not. R. Astr. Soc.* **198** 947–961.

Schofield, N. and Woolfson, N. M. (1982b) *Mon. Not. R. Astr. Soc.* **198** 963–973.

Seaton, M. J. (1955) *Annls. Astrophys.* **18** 188–205.

Spitzer, L. (1939) *Astrophys. J.* **90** 675–688.

Stephenson, A., Runcorn, S. K., and Collinson, D. W. (1975) *Geochim. Cosmochim. Acta Suppl.* **6** 3049–3062.

Stock, J. D. R. and Woolfson, M. M. (1983a) *Mon Not. R. Astr. Soc.* **202** 287–291.

Stock, J. D. R. and Woolfson, M. M. (1983b) *Mon. Not. R. Astr. Soc.* **202** 511–530.

Toksöz, M. N. and Solomon, S. C. (1973) *Moon* **7** 251–278.

Tyror, J. G. (1957) *Mon. Not. R. Astr. Soc.* **117** 370–379.

von Weizsäcker, C. F. (1944) *Z. Astrophys.* **22** 319–355.

Williams, I. P. and Cremin, A. W. (1969) *Mon. Not. R. Astr. Soc.* **144** 359–373.

Williams, S. and Woolfson, M. M. (1983) *Mon. Not. R. Astr. Soc.* **204** 853–863.

Wise, D. U., Golombeck, M. P., and McGill, G. E. (1979) *Icarus* **38** 456–472.

Woolfson, M. M. (1964) *Proc. R. Soc. Lond.* **A282** 485–507.

Woolfson, M. M. (1979) *Phil. Trans. R. Soc. Lond. A.* **291** 219–252.

Woolfson. M. M. (1984) *Phil. Trans. R. Soc. Lond. A.* **313** 5–18.

Yabushita, S. (1979) *Proc. IAU Symp. No. 81, Tokyo.* Reidel: Dordrecht.

Yabushita, S., Hasegawa, I., and Kobayashi, K. (1982) *Mon. Not. R. Astr. Soc.* **200** 661–671.

Plates

Plate 1 — The rings of Saturn as seen by Voyager I. (Courtesy of Jet Propulsion Laboratory.)

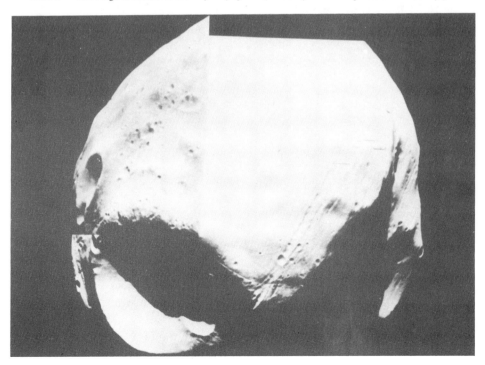

Plate 2 — The Martian satellite Phobos. (Courtesy of Jet Propulsion Laboratory.)

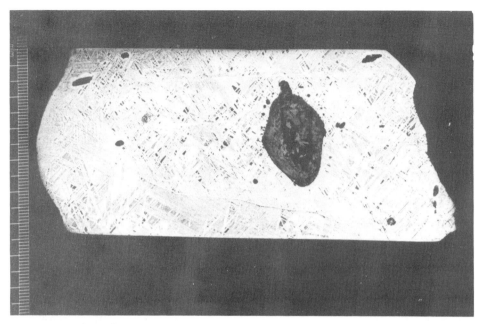

Plate 3 — A polished and etched section of the iron meteorite Gibeon. The Widmanstätten pattern is clearly seen. The dark regions are troilite, a sulphide of iron. (Provided by the British Museum.)

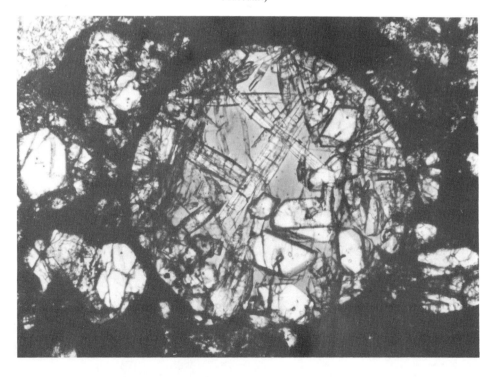

Plate 4 — A well-formed chondrule (diameter approximately 1 mm) in a chondritic meteorite. (Courtesy of Dr R. Hutchison.)

Plate 5 — A moon globe showing the face seen from earth.

Plate 6 — A moon globe showing the reverse face.

Index